Rx: Coconuts!
(The Perfect Health Nut)

Rx: **Coconuts!**
(The Perfect Health Nut)

How the Coconut Can Enhance Your
Life and Well-being

Vermén M. Verallo-Rowell, M.D.

Copyright © 2005 by Vermén M. Verallo-Rowell, M.D.

Library of Congress Number: 2005903595
ISBN :   Hardcover   1-4134-9387-4
          Softcover   1-4134-9386-6

All rights reserved. No part of this book may be reproduced or transmitted in any form or by any means, electronic or mechanical, including photocopying, recording, or by any information storage and retrieval system, without permission in writing from the copyright owner.

This book was printed in the United States of America.

To order additional copies of this book, contact:
Xlibris Corporation
1-888-795-4274
www.Xlibris.com
Orders@Xlibris.com
28455

CONTENTS

INTRODUCTION 11
        How this Dermatologist Fell in Love with the Perfect Health Nut

PART ONE: WHY THE COCONUT IS THE PERFECT HEALTH NUT

Chapter 1:   Coconut Oil Will Help You Lose Weight 25
Chapter 2:   The Skinny on Saturated Fat . . .
                 It Can Be Good For You 39
Chapter 3:   Coconut Oil Is Cholesterol Free,
                 Trans Fat Free, and Heart-Healthy 53
Chapter 4:   Coconut Oil May Help Protect Against Cancer 69
Chapter 5:   Coconut Monoglycerides:
                 Impressive Results in Infectious Diseases 91
Chapter 6:   Far From Giving You Acne, a Coconut Oil
                 Derivative Can Help Treat It 111
Chapter 7:   Virgin Coconut Oil Is More Than Just an
                 Ideal Skin Moisturizer 121
Chapter 8:   Coconut Water Growth Factors:
                 Can Potentially Retard Aging to
                 Rejuvenate Hair and Skin 133

PART TWO: A DOCTOR'S PRESCRIPTION FOR USING THE COCONUT

Chapter 9:   Rx Number 1   How to Use Coconut Oil,
                                   Water and its Monoglycerides for
                                   Internal and External Health 149
Chapter 10: Rx Number 2   How to Select the Right Coconut Oil,
                                   Or "The Tale of the Two Virgins,
                                   Coco And Olive" 161

Chapter 11:	Rx Number 3	The Rx: Coconuts Lifestyle…A Prescription for Head-to-toe, Cell-to-surface Health and Beauty	... 177
Chapter12:	Rx Number 4	Comparative Nutrition Guide and Tasty Recipes	 197

DIRECTORY OF COCONUT PRODUCTS 231
REFERENCES 245

Dedication

To my father, Felipe R. Verallo,
the gentle farmer from Cebu and Leyte,
together with Laura, CC, and Glendon...this is for you, Pa.

Acknowledgements

This book became a reality only because of the generosity of my family. Without Glendon's perspectives, comments, suggestions, first-hand editing, and CC's loving attentiveness, this would not have been written. Special mention is made of Laura, semiotician, and English major . . . all-around editor, organizer, co-writer, graphic designer, book illustrator, she made the book come to life in this its final form . . . I love you all, and thank you very much. To my son-in-law, Juan Pablo: my appreciation for missing your beloved futbol to provide invaluable help, general scurrying, and support during the final, very long days of writing!

Dr. Conrado Dayrit, my dear Professor, thank you for sharing with me your vast knowledge base and wisdom.

Dr. Jon Kabara, thank you for your support and inspiration in pursuing this book and the monolaurin clinical studies.

Dr. Florentino Solon and Dr. Rudolfo Florentino, for your perspectives on nutrition, and Dr. Teresita Espino for introducing me to a different kind of coconut oil/derivative to study.

Many thanks to: Philippine Coconut Authority Past Director Evangeline Valbuena, for her advice and assistance; Mr. Sing Tiu, and Engineer Divina Bawalan, for showing to me the practicalities of the coconut world.

To Todd Manza, my relentlessly fastidious yet patient Copy Editor

Chefs Lisa Alvendia, Pia Castillo-Lim, Beth Romualdez, Myrna Segismundo, and Dietician Ms. Imelda Peralta, it was a pleasure working, and tasting, the coconut recipes with you.

Author's photograph by Joanne Zapanta-Andrada: thank you, Jo, for always making me feel at ease in front of the camera.

To my residents at the Philippines' Makati Medical Center and the Skin and Cancer Foundation: invaluable and brilliant as always. I look forward to watching your careers shoot up as high as the tallest coconut tree!

To my patients: thank you for being willing participants and for your feedback. Truly, without you, this book could not have gotten off the ground. And as you know all too well, I share your excitement at how great you look and feel today!

To my new friends, villagers and farmers who respect the land and the coconut . . . Mabuhay.

Last and equally important, a special thank you to Francis J. Ricciardone, the United States Ambassador to the Philippines, who, from his unique vantage point as a multicultural diplomat, questioned and assessed the science, then helped me achieve balance in my proposals for a coconut lifestyle.

Introduction

How this Dermatologist Fell in Love With the Perfect Health Nut

The study of skin is my passion. This obsession started at the Cleveland Clinic, where I trained in Dermatology during the sixties. After completion of my basic studies came the hunger to learn more, to dig deeper into the look and the lives of skin cells. So I went on to study the subspecialty of Dermatopathology with Dr. Hermann Pinkus in Detroit.

These two disciplines made me incurably curious about the skin and how it works. I learned to observe and to question everything. When the answers were dubious or indefinite, I started to undertake research projects to get a more exact answer. At first my studies were simple case reports of an unusual disease or the rare kind of patient. When I started my own specialty practice, I learned that I could apply the results of those research studies to innovate, initiate, and probe into newer and better treatment options for the benefit of my patients.

Through these studies I have had the opportunity to lecture at international dermatological and medical forums, presenting my research findings. These studies have been published in peer-reviewed medical journals, books, and conference abstracts, and have been discussed extensively with colleagues with mutual interests. The results of some of the studies have even resulted in patented products that now have stood the test of time.

But . . . none of these studies were about the coconut.

Why then—after 30+ years of doing conventional medicine studies—am I writing about the coconut? This subject seems to belong to alternative medicine and has never really had a place in mainstream medicine. What was the impetus for me to climb up this tropical tree of knowledge?

The reason is threefold: my insatiable curiosity, my exposure to different cultures and ways of thinking, and my roots. First, my curiosity: besides being innate in me, it was, as I mentioned, honed, buffed, and polished during my residency, and has steadily grown throughout my life and my practice. Second, my exposure to different cultures has a long history, both personal and professional. I trained in many parts of the United States and have lectured in countless countries around the globe. My husband is from New England and we keep a home in a picturesque, seaside town in Maine, as well as in Hong Kong and in mainland China, where he does business. One of my two daughters works in New York and the other is married to an Argentinean, whose home we have visited often.

Third, my roots have always been important to me. I grew up in Cebu, an idyllic tropical island, and I return often for personal and even professional reasons (such as medical outreaches). There, the coconut tree grows everywhere and has a starring role in much of our folklore. For example, folklore attributes the beauty of thick, shiny, jet-black hair and the gorgeous brown skin of our islanders to the use of coconuts. Even though exposed to outdoor sun all their lives, our farmers and fishermen have fewer wrinkles than the Caucasian tourists who come to visit. They also hardly ever develop skin cancer.

This is, actually, what first caught my attention and led to me doing a study on the sunscreen factors of brown skin, which I then published in my first book, *Skin In The Tropics: Sunscreens and Hyperpigmentations*. What I discovered was that the sunscreen factor of brown skin, even after it was darkened from exposure to the sun, is SPF 2 to 4 *at most*. So there had to be another reason why the cancers and even the precancers of the skin rarely developed among our island men and women. Was it a simple matter of luck, a fortuitous draw of the genetic gene pool? Were they just biologically less prone to skin cancers?

More and more of the medical research I had been reading began to emphasize the importance of diet in the promotion or prevention of cancers . . . could diet, therefore, have something to do with their skin's natural resilience to cancers? Later, it occurred to me . . . almost all folklore has some element of truth. Was it possible? . . .

Could the coconut play a part?

Then, in 1998, I had a fortuitous meeting with Dr. Jon Kabara at a dermatology convention in Beijing. Dr. Kabara, a Ph.D. in Pharmacology, holds several patents on fat and oil products. He is one of a number of academic researchers, mostly from the United States and some from Europe, who since the 1960s have been undertaking studies of dietary fats and oils, including coconut oil and the chemicals derived from it. His work introduced me to a new perspective on the coconut, particularly his discovery of the impressive antiseptic properties of monolaurin, which I discuss further on. At that time, my talks with Dr. Kabara prompted my medical interest in and some initial studies of the coconut—but, for the moment at least, these were limited to monolaurin's potential as an antiseptic and a nonallergenic preservative for cosmetics (the results of which follow, below).

In the meantime, I focused on my first book, an explanation of how the sun and different forms of light affect skin using various studies I had published over the years regarding light-induced skin problems like cancers and hyperpigmentations. As is not uncommon for me, the process of writing that book led to new curiosities, new topics for further research. In addition, as I continued with my clinical practice and research, I started getting more questions from my patients about botanical products. I encountered more papers beginning to look into the potential of botanicals not just for the skin but also for our diets and our overall health. I also noticed that in recent years botanicals had become a rich source of dermatological products, with many claims about what one or another plant extract could actively do for the skin. I knew my curiosity was growing, and that I would eventually do studies about these botanicals . . . I just hadn't pegged down exactly which plant or what I wanted to discover.

Taking a break from writing one day, I sat in our garden to enjoy the lush vegetation and the riot of colors from our tropical flowers. I happened to look up, and my eyes rested on our stately coconut trees, capped by heavy bunches of perfectly formed coconuts. My book was in my head—as was the natural resistance to skin cancer of the brown-skinned subjects I mentioned earlier, as was this growing interest in botanicals, as was the ever-increasing body of research from Dr. Kabara's team. "Focus," I remember being told by my mentor, Dr. John Haserick, when I was an over-eager resident. At that moment, it hit me! I decided, then and there: my next studies would focus on not just any botanical; I would focus on *this one*, the coconut.

Coconut oil is rich in a chemical called lauric acid, which in the intestines naturally becomes a substance called monolaurin. In 1972, Dr. Kabara first described the significant antiseptic properties of monolaurin (197), an observation that has been validated and further expanded by many other researchers (210-219). This finding should have been a landmark discovery, much like the discovery that penicillin can be synthesized from the fungus *Penicillium*. However, monolaurin was relegated to complementary and alternative, rather than mainstream, medicine. Even today, a surprising number of doctors have never heard of monolaurin.

Armed with the knowledge gained from my discussions with Dr. Kabara and my recollection of the curiously good appearance (and cancer resistance) of the skin of farmers and fishermen in the tropics despite their exposure to the sun and the sea, I initiated dermatological studies using modern methods that follow a body of procedural rules. Currently, this is considered the most reliable and meaningful medical research, popularly called Evidence-Based Medicine.

Curious, but skeptical, I decided to start with a simple study. Alcohol—besides being in your favorite cocktail—is also *the* time-tested antiseptic. Using a hand gel glove-dip testing method of the U.S. Food and Drug Administration (FDA), I first studied how monolaurin would compare with alcohol as an antiseptic. The results were promising: a gel we prepared with only 1.5% monolaurin was *as effective* as a 70% isopropyl alcohol gel in eliminating the test bacteria immediately; and after 3,5,7, and 10 glove-dips (223).

The following year (1999), I decided that those results merited a follow-up with a more substantial, randomized, double-blind clinical study. This kind of study is respected as more scientifically valid because no one—neither the participants nor the clinicians—knows which samples are involved until the tests are complete and a statistical analysis has been done.

The subjects of our study were nurses coming off their shifts at a busy hospital (224). The nurses were asked to not wash or to apply any antiseptic to their hands after their shift. First, we dipped the nurses' unwashed bare hands into gloves filled with water. A microbiological lab then cultured this water to determine which organisms the nurses tended to pick up from their day's work.

Once this was determined, measured amounts of the top six organisms that grew and flourished on hands exposed to a hospital environment were rubbed on the nurses' hands. Immediately after, the nurses were divided into three groups and asked to rub their hands with three different "antiseptic gels." Without knowing which antiseptic gel they were getting, Group 1 rubbed their hands with monolaurin, Group 2 rubbed their hands with a 70% alcohol hand gel, and Group 3 rubbed their hands with a simple saline solution as a control. After this, all the nurses dipped their hands into another glove of water. Again, the water was cultured to determine how many of the organisms were destroyed and prevented from growing. This was done within 30 seconds, and 5 minutes after the hand gel application.

As expected, the nurses whose antiseptic gel was just the saline solution grew the organisms, and the nurses who used the 70% alcohol hand gel eliminated all the cultured hand bacteria. To our amazement, the nurses who used the gel containing 1.5% monolaurin also eliminated all the cultured hand bacteria. The statistical analysis validated these results.

As a dermatologist, I wanted to know what other effects these natural coconut derivatives could have on the skin. More evidence-based studies followed. Some studies continued to focus on the antiseptic activity of monolaurin on several other kinds of bacteria (226-227). I also tested monolaurin on fungus (246-248) and on viral organisms (225) that produce skin infections. I even tested it as an antiseptic ingredient in cosmetic and skin care products.

The results were truly surprising: monolaurin from the coconut was as good in its broad-spectrum antiseptic effects on human skin as it had been in Dr. Kabara's cultures of microorganisms in the laboratory (198-209).

On the famous, ultrawhite coral beaches of the Philippine island of Boracay, as in my home island of Cebu, the local people use coconut oil on their skin and when giving massages to tourists—who never seem to get enough of it. I therefore conducted another double-blind study on the oil's effects as a moisturizer, comparing it with mineral oil, which is commonly used either as-is or as an ingredient in other products to moisturize the skin (249). Coconut oil proved as effective as mineral oil, *plus* it possessed the potential broad-spectrum antiseptic effects we had observed from monolaurin. Then we determined that coconut oil protects the skin from the sun, as an antioxidant (262).

The results of these studies were such eye-openers that I started to wonder about the coconut's other benefits, benefits beyond dermatology. If all of these things I was discovering were still unknown to most dermatologists, was it possible that there were also food facts still to be discovered about the coconut? I have always had a dermatologist's professional interest in finding a medically sound diet that will make my patients leaner and healthier. A leaner and more energetic body generally goes together with a healthy, more youthful look, and with glowing skin. Health problems can lead to skin problems as well—on their own or as reactions to the medicines used to treat them.

On a personal note, I have always been interested in nutrition, being petite and hyperconscious of what additional weight could do to my look. I have tried almost every diet (even as early as the 1960s, during my training at Cleveland Clinic, I joined several clinical studies on cholesterol), searching not only for the most effective, but also for the healthiest, easiest, and most pleasant option. Since then, I've seen the best (and happiest) results from the Atkins diet for a quick fix, the South Beach diet for a less painful long run—modified with the coconut, in a "lifestyle" that I'll share with you in Chapter 11—and a good combination of sustained cardio and light weights, thanks to my drill-sergeant of a personal trainer!

During this long search, I realized that, like many of my medical doctor colleagues, I only had a rather basic understanding of nutrition. To remedy this deficiency, I studied the related research materials—only to realize that many of the nutritional experts who had been dictating how we should eat were giving out advice that has been convincingly challenged. For instance, there was the diet-heart theory that emerged in the 1960s, giving rise in the 1980s to the dietary recommendations of the American Heart Association. That, in turn, became the basis for the U.S. Department of Agriculture's (USDA) Food Guide Pyramid—which, thanks in part to the research of Drs. Atkins and Agatston, has since been called into serious question.

Like these doctors, I consider the food pyramid "flawed" because we have substantial and all-too-apparent proof that its dietary recommendations have not worked. This nutritional scheme has resulted in more obese and even ultra-obese people than ever before. Incidence of other problems related to the food we eat—diabetes, hypertension, and heart disease—have also reached an all-time high. Despite our adherence to the diet-heart theory that was supposed to make us heart-healthy, heart disease ranks among the leading causes of death, worldwide (52).

My search eventually led me to the little-known nutritional research done by professors from the tropics. Their scientific curiosity had focused on the oil from the fruit of a tree that is abundant where they live, the coconut, and they had made some amazing discoveries. They were eager to share their good findings, and they published them, but no one was interested—including me, until I belatedly tuned in. Now, encouraged by the surprising results of my own studies on coconut oil and the skin, I decided to take a longer, more serious look at their research. As I reviewed these studies, I found the facts almost too good to be true. Yet the results of their studies were consistent, and the proof was scientifically sound. The evidence overwhelmingly illustrated that, at worst, the coconut is *neutral* to the heart, and at best, the coconut is actually *good* for the heart (60-77). I saw this in the various studies on laboratory animals (73), in clinical studies using human participants, and in studies of whole populations in coconut-eating countries.

Other studies showed that whether over short or long periods of time, people who were ingesting coconuts daily were heart-healthier than those who ate other kinds of fats and oil. The non-coconut-eating world was definitely missing something!

On my journey to learn more about the oils in our diet, I stumbled onto something even more intriguing: whereas certain kinds of fats promote the development of *cancer*, coconut oil does not (98). Now I was really determined to know as much as I could about the coconut and all its uses! It became more obvious to me, from those studies and from the results of my newer research, how wrong I was to have dismissed the coconut for so long. In fact, it began to dawn on me that coconut oil might even be the *best* of all the oils we eat and use: it is impressively healthy for the heart, helps prevent disease, serves as a natural treatment for infections, and keeps skin looking youthful, clear and beautiful. An all-around health nut!

Why, then, did I ignore this great oil for so long? And I wasn't the only one . . . . Why, if all these studies were showing such fantastic results, had coconut oil been generally forgotten or neglected or even maligned as a harmful dietary oil? The simple answer is that coconut oil is a saturated oil and there is a firmly ingrained assumption that all saturated fats are the same: rich in cholesterol and a cause of heart disease. As the studies and discussions in this book will show, however, this assumption is wrong. Although saturated fats from animal sources and dairy products are cholesterol rich, for example, *coconut oil is derived from a plant and is in fact cholesterol free!*

Part of the reason this fact is not more widely known is an economic one. Coconut oil used to be a staple in many more countries, including in the United States. However, during the 1960s, the U.S. domestic vegetable and seed oil industry promoted their oils by emphasizing only half the truth about coconut oil. They mentioned that it is a saturate, and encouraged the blanket assertion that saturated equals bad, but they neglected to mention that coconut oil is also cholesterol free.

In the 1980s, a powerful activist group called the Center for Science in the Public Interest (CSPI) publicized this half-truth with a high-profile media campaign. Founded in 1972, and later revealed to have been supported by the soybean industry, the CSPI's major campaign against the use of saturated fats for frying was launched in 1984 (50).

In 1986, the CSPI added coconut oil to its list of saturated fat "villains." In their publications, which included full-page newspaper ads, they condemned the coconut as hydrogenated and cholesterol rich. This was no longer a half-truth; it was wholly untrue. Coconut is a stable oil, does not need to be hydrogenated, and is cholesterol free (50).

In 1988, the CSPI published a booklet called *Saturated Fat Attack*. Their message was that all "tropical oils" should be banished from food. This boosted the use of soybean and the other vegetable and seed oils while virtually eliminating coconut oil from the American diet and from most other diets in the developed world. The CSPI could have had an economic agenda—remember that they were supported by the soybean industry, which their work catapulted into market dominance. Even if they had the best of intentions, however, and really believed that all saturated "tropical oils" posed a health threat, the science was incomplete, the information skewed, and the conclusions erroneous (50).

Sadly, most people, even physicians, did not question these media pronouncements. Worse, researchers became reluctant to perform studies on an oil that apparently, through the power of industry-supported media, had now become anathema. In an academic culture, researchers are expected to "publish or perish," and getting any research paper published is quite difficult. A paper based on a highly unpopular thesis suffers a greater likelihood of never seeing print or of being accepted only by journals or books with limited circulation. It is therefore understandable that for most researchers through the years the obvious choice was *not* to study the coconut. Only a few professional investigators were willing to risk their own reputations to defend or even to study coconut oil. So, in the absence of any prestigious counterclaims, the public believed the press, and today coconut oil has been practically eliminated as a U.S. food ingredient.

The proof of coconut oil's benefits, however, is just too powerful to ignore, and I have decided to join the ranks of the likes of Dr. Kabara, and Dr. Dayrit, not only to defend the coconut but also to extol its virtues. I have switched to the "Coconut Lifestyle," as have my children, my husband and scores of my patients and friends—and we are healthier (and better looking!) for it. And as this book will show, our satisfaction is medically sound and scientifically supported.

In recent years, two physicians have written hugely popular nutrition books: Dr. Robert Atkins's *New Diet Revolution* and Dr. Arthur Agatston's *The South Beach Diet*. Why are they popular? Because they produce results! Drs. Atkins and Agatston emphasize, with some variation, that the best diet is one that is low in carbohydrates and high in proteins. Both doctors are physicians who, like me, are *not* nutritionists. We are practicing M.D.s seeking a diet that works for our patients and for ourselves.

As important and impressive as their works are—I, myself, continue to live by their tenets—neither doctor mentions the coconut. This is a shame because coconut oil works beautifully with all diets, especially theirs. In order to show you how, I will help you focus on the fat part of your diet. In this book, you will learn the basics about the different kinds of fats and oils—we'll whittle it down and decode all the mumbo jumbo—to really determine what is good, what is bad, and *why*. Instead of blindly following on the old mantra, "eat *low* fat," you will learn to take control by finding out which fats you can eat without feeling guilty or irresponsible . . . even which fats you *should* eat for better health. Your mantra can now be "eat *good* fat."

Backed by medical and scientific facts, I'll take you on the same journey of discovery that made me fall in love with the coconut and that led me to the conclusion that the coconut is the Perfect Health Nut. You'll learn how the coconut can help you avoid obesity, how it can boost your immunity, and how it can protect you from bacteria, fungus, and viruses—all while keeping you heart-healthy and moisturizing your skin to a natural glow and beauty, even treating acne, and providing beneficial antiseptic, tumor-protecting, and antioxidant effects.

It gives me great pleasure to recognize that all these attributes I can also offer to my friends whose religion requires their food to be prepared in a special manner. These coconuts are naturally **halal** for my Muslim friends, and **kosher** for those of the Jewish faith.

Rx: Coconuts! (The Perfect Health Nut)

To get everyone started on the "Rx: Coconuts Lifestyle," I've included tasty, coconut-focused recipes from stellar Asian chefs and a product guide of available coconut products.

In a nutshell, this book is a doctor's prescription for a diet and lifestyle that keeps you heart-fit, slim, disease free, more cancer resistant and gorgeous. In essence, it is a prescription for a way of being, one that rewards you with a lifetime of true head-to-toe (and even cell-to-surface) health. In this Rx, simply live the coconut lifestyle!

Part One

A DOCTOR'S PRESCRIPTION FOR USING THE COCONUT

Chapter One

Coconut Oil Will Help You Lose Weight

In This Chapter

- Coconut Oil is Made Up of Mostly Medium-chain Fatty Acids
- Coconut Oil is Low in Calories
- Coconut Oil Has Good "Mouth Feel"
- Recent Studies: Coconut Oil Increases Energy Expenditure and Satiety
- Selection Guide: A List of Long-chain Oils and Medium-chain Oils
- In a Nutshell

Y ou read right! The regular intake of coconut oil can actually help you lose weight. How? Primarily by being a medium-chain oil. In this chapter, we'll see how, even a simple act of replacement—reducing long-chain oils in your diet, and replacing them with medium-chain oils like coconut oil—can help you lose more weight healthily, and keep it off. If you're already happy with your diet, this simple switch means you can stick to it and get better results. On the other hand, if you're still looking for the perfect weight-loss regimen, switching to coconut oil can jumpstart your slimming process.

It is important to remember, as with any proper weight-loss regimen, that significant and sustained weight loss entails a balanced diet and regular exercise—in other words, switching to coconut oil alone while gorging yourself on donuts and shunning the treadmill will not help you lose weight. What I am proposing here is that because of coconut oil's unique properties, it can replace the oil in any weight-loss diet, enhance that diet's results, and reward you with the other benefits detailed later in this book. In short, this section will show you that simply by substituting your regular oil with coconut oil, you can help maximize your particular diet's ability to help you lose weight and keep it off.

Coconut Oil is Made Up of Mostly Medium-chain Fatty Acids

T he best way to understand how coconut oil can make you leaner is to have a better understanding about the fats and oils in your diet. Admittedly, this can be intimidating, which is why I have decided to focus first just on fatty acids. With a better understanding of fatty acids, you already will be empowered to make the right choices regarding the fats and oils in your diet.

All fats and oil are made up of chemical compounds called triglycerides. A triglyceride is made up of glycerol, a molecule made up of three carbon atoms. Each of these three carbon atoms has one fatty acid attached, for three fatty acids in total (hence the "tri"). In turn, these fatty acids are composed of chains of carbon atoms. Depending on the length of the carbon chains, these fatty acids are called *short chain, medium chain,* or *long chain*. The length of the carbon chains is determined by the number of carbons in the chain. The long chains for example are C_{14}, C_{16}, C_{18}, C_{20}, C_{22}, and C_{24}, and the medium chains are C_8, C_{10}, and C_{12}. This information will come in handy later on, but for now, all you need to remember is that the fatty acids that make up coconut oil are mostly medium chain.

The majority of fatty acids in the food we eat are either medium-chain or long-chain molecules (they have medium or long carbon chains). Just like tall people who can easily put their carry-on suitcase in the luggage rack above an airplane seat, short people like me just can't. It's a simple matter of physical difference. Long-chain fatty acids are also physically different from medium-chain fatty acids. Foods that are made up of mostly medium-chain fatty acids are shorter and smaller, are absorbed easier, are used up faster, and have fewer calories per gram. Therefore, less of their fatty acids become stored in fatty tissues than do those of foods composed of long-chain fatty acids. Long chains are physically longer so they take longer to get through your digestive system, have more elements to break down for energy use, and increase the risk of being stored as fat.

OIL, FAT, AND ENERGY: 101

Following an Oil with Long-Chain Fatty Acids through Your System. Let me give you a simple demonstration of how oils and fats are generally handled by and work for your body. To do this, we follow a drop of oil from a juicy slice of steak through the process of digestion, absorption, and conversion to energy. Our oil in this example is steak oil, a triglyceride with mostly long-chain fatty acids. As the oil goes from your mouth to your stomach, the process of digestion breaks it down by physical and chemical means into smaller and smaller bits and pieces. Just about the time the taste of the steak is a delicious memory, the oily bits are ready to go into the intestines. There, these oily pieces continue to break down in the presence of enzymes called lipases and under the influence of bile—a substance that may sound ugly but that is actually instrumental for our digestive system to function as well as it does.

Under the influence of these chemicals, the fat particles in the oily bits continue to break down into smaller components. These components are called by names that indicate their size: the steak oil breaks down from its original *triglyceride* state to its *diglycerides* and then to its *monoglycerides*. At the very end of the digestion process, these glycerides are further reduced to any oil's smallest chemical units: fatty acids and glycerol.

After digestion, these fatty acids are now ready for the next step, absorption. Tiny as the fatty acid and glycerol are, they now have to contend with yet another physical barrier to get through your system: the intestinal wall. The intestinal wall allows water through it, but not fat. What does your body do? The cells of the intestines conveniently form *phospholipids*, compounds that are made up of both a fat and a protein portion. The fatty acids wrap around the fat portion of the phospholipids while the protein portion, which contains water, gets them to the other side of the intestinal wall. In our steak oil example, its long-chain fatty acid compounds arrive on the other side of the intestinal wall as triglycerides again.

As you can see, nature is marvelously efficient in handling your triglycerides. Imagine sending a box of chocolates from New York to Los Angeles using, say, FedEx. If FedEx operated like the human body, the company would process the chocolates en route into a small bundle of its component milk, cocoa, and sugar, and would then restore these to the original chocolate form and packaging by the time they reached L.A.

Like the repackaged chocolates on the FedEx truck, your steak oil's long-chain triglycerides re-form as they leave the intestines. From there, they travel in tiny circulatory vessels called lymphatics, which carry a liquid called lymph. These lymphatic vessels eventually get to the largest lymphatic vessels, which in turn empty into the vessels that carry blood. This is why the level of triglycerides in your system can be measured by a blood test.

Inside these blood vessels, the long-chain triglycerides circulate in the body and finally reach the liver. There, the long-chain triglycerides are broken apart once again into their basic fatty acid and glycerol components.

Still tracking that drop of steak oil, we now go to how they contribute to your body's energy. At the level of your cells, these fatty acid and glycerol components go through a process called beta-oxidation, a basic process of life that uses the oxygen you constantly breathe. During beta-oxidation, enzymes force the fatty acids to break up even further into small two-carbon subunits that your cells can use. When energy is needed, these two-carbon subunits release adenosine triphosphate (ATP), the brand name for the gasoline in your body's engine.

ATP is the basic unit or molecule you use for all your energy needs. If the food you eat and your activities are balanced, the fatty acids from your diet are *all* broken up into the ATP energy units and *all* are used up. When you eat more than you need, the fatty acids are present in excess, and recombine to form triglycerides again, which accumulate as fat for long-term parking.

To make it simple: eating more than you need >> too many fatty acids that cannot be broken down and converted into ATP >> the excess fatty acids recombining into triglycerides >> long-term fat parking >> the fat rolls and deposits many of us see in the abdomen, buttocks, and thighs.

The process I have just described is the digestion and energy production of the triglyceride fatty acids, which are long chain and therefore larger. These long-chain fatty acids are the ones present in vegetable or seed oils like soybean and corn, and in animal fats and dairy products.

Medium-chain Magic. Although coconut oil is derived from a plant—like the vegetable or seed oils mentioned above—nature made the fatty acids of its oil uniquely different from all these other oils. Coconut oil's fatty acids are mostly medium chain. The difference between the medium-chain character of coconut oil and the long-chain character of other vegetable, seed or animal oils is critical to our understanding of why coconut oil is a less fattening oil.

The physical difference in length is the first reason for why coconut oil is less fattening than the other oils in your diet: the absorption of coconut oil's mostly medium-chain fatty acids is different from the absorption of the mostly long-chain fatty acids in other vegetable or seed oils, and in the steak oil example we discussed above.

The size of the medium-chain fatty acids of coconut oil is smaller than that of the long-chain fatty acids. This allows them to go through the digestion, absorption, and energy conversion described above much more quickly than the long-chain ones. On the inside of the intestines, the smaller size of the medium-chain fatty acids allows them to get directly inside capillaries, the smallest vessels that carry blood. At the other side of the intestinal wall, they do not arrive at vessels that carry lymph. Instead, from the capillaries they proceed into larger blood vessels, and shortly thereafter they reach the large-portal blood vessels that serve as a gateway from the intestines into the liver.

After this shorter route, the medium-chain fatty acids of the coconut can more rapidly be broken down into those ATP energy molecules. In the steak oil example, on the other hand, we saw how the long-chain fatty acids follow a long, circuitous route. They first circulate in the body and contribute to fat deposits, before ending up in the liver.

Like the ease with which your smaller compact car, compared to a limo, is able to get in and out of New York's traffic, the medium-chain fats take the shortest route to their destination, and have little opportunity to create a traffic jam within the blood vessels. In contrast, the long-chain fatty acids circulate more, are retained in your body's fat deposits and have more opportunity to adhere to the walls of your blood vessels, leading to a potentially catastrophic "traffic jam": a closed heart vessel or a stroke.

This physical size difference is also the reason for coconut oil's lower calorie count.

Coconut Oil is Low in Calories

This is one delicious secret: *coconut oil is lower in calories than any other oil in your diet.* Upon reading that, you may have had a gut reaction of "low calorie = great!" But I wrote this book to give the average reader a better understanding of what all the hip, hyped keywords of fat loss mean. So that you may fully appreciate coconut oil's low-calorie character, allow me to review with you just what *calories* mean.

Just as logs of wood are burned in a fireplace to give out heat, the food you eat is burned in your cells, to produce the energy which allows you to work out at the gym or just to go about your daily, ordinary errands. You count the number of logs that, when burned, can give off enough heat to keep you warm on a cold winter night. With food, you count the amount of food, in grams, that produces the calories you need for any activity.

"Calories per gram" is the measure for anything you eat, regardless of which food group it belongs to—proteins, carbohydrates or fats. Proteins are abundant in steak, fish and chicken. They produce *four* calories per gram. Carbohydrates are the primary food source for quick energy. They provide *four* calories for each gram of carbohydrate in the food you eat. You get these carbohydrates mostly from potatoes, from rice, from bread, pasta, and anything else made of flour, and from the sugar in your drinks and desserts.

The fats and oils in food, also called *lipids*, yield nine calories per gram. That is, there are *nine* calories per gram in oily food versus the *four* in carbohydrates or protein. Now you understand why, in general, you need to exercise twice as much to burn a gram of fat from the steak you ate as you do to burn a gram of its protein or the carbohydrate from the dessert that followed.

***But*—Not All Fats and Oils Are the Same . . . Coconut Oil Actually Produces Fewer Calories.** Earlier in this chapter you learned that the fatty acids of coconut oil are different from those of the vegetable or seed oils because they are medium chain. As such, coconut oil is absorbed faster than the long-chain oils (remember that this was the first reason why coconut oil is less fattening). This difference in chain lengths also provides the second reason why coconut oil is less fattening than other oils: coconut oil produces fewer calories than long-chain fatty acids.

Nutrition experts have written about this difference since the 1950s (1). Their studies showed that medium-chain fatty acids such as those in coconut oil produce fewer calories than the long-chain fatty acids from vegetable or seed oils. This difference is again based, in part, on physical size: the medium-chain fatty acids are shorter and therefore produce fewer two-carbon, energy-producing molecules; the long-chain fatty acids produce more. Additional lipid studies, first in animals and later in humans, have confirmed the existence of this difference (2-6).

Let me demonstrate this difference using a product called MCT oil. This commercially prepared oil is made up of medium-chain fatty acids from coconut oil. The effective energy value of MCT oil is not 9 calories (like other oils) but just *6.8 calories per gram* (7). Besides the lower calories, the medium-chain oils make MCT oil easier to absorb. It has therefore been used for many years in nutritional and medical formulas—as a food for weak or premature infants, burn patients, and the elderly—since these patients obviously need a food that is easy to digest and to convert into energy.

Here's where you can see how the length of the carbon chain makes a difference in the number of calories per gram of an oil. MCT oil has about 75% C_8 and 25% C_{10}. The number of calories is 3,084 per pound.

Coconut oil has about 8% C_8, 7% C_{10}, and 50% C_{12}. The number of calories is 3,910 per pound. Note that MCT oil is 100% medium-chain triglycerides whereas coconut oil is about 65% medium-chain triglycerides; hence, MCT oil's much fewer number of calories per pound.

Now compare this to the fatty acids in vegetable or seed oils, which are all long chain and which add up to about 4,010 calories per pound (7). This is a one hundred-calorie-per-pound difference, or 2.5% fewer calories in the medium-chain fatty acids of coconut oil than in the long-chain fatty acids of the vegetable or seed oils—including even my other favorite, olive oil. (But this is balanced by the many positive things about olive oil that we'll talk about later in the book).

Want to be further impressed? Let's look at the following three tables that show the Nutrition Facts of different dietary and coconut products. The first table compares coconut oil to olive, canola, soybean, safflower and fish oils from salmon. For the same amount of these oils, coconut oil has the least number of calories.

CALORIE COMPARISON: VIRGIN COCONUT AND OTHER DIETARY OILS

	SIZE	CALORIES	CALORIES FROM FAT
COCONUT	1 cup (218 g)	1879	1879
OLIVE	1 cup (216 g)	1909	1909
CANOLA	1 cup (218 g)	1927	1927
SOYBEAN: SALAD OR COOKING	1 cup (218 g)	1927	1927
CORN: SALAD OR COOKING	1 cup (218 g)	1927	1927
SAFFLOWER	1 cup (218 g)	1927	1927
FISH, SALMON	1 cup (218 g)	1966	1962

Source: www.Calorie-Count.com, May 2005.

The difference in calories is even more when we compare coconut milk and cream with dairy butter in the next table.

CALORIE COMPARISON: COCONUT MILK AND DAIRY BUTTER

	Coconut Milk Coconut Cream	Dairy Butter	You can have substantially less calories by adding coconut milk or cream and lessening butter.
Size	1 cup (240 g)	1 cup (227 g)	
Calories	552	1628	
From Fat	515	1628	

Source: www.Calorie-Count.com, May 2005.

In this chapter on weight loss, we're focusing primarily on coconut oil. However, allow me to digress slightly to share with you some more information about how another coconut product—its juice or water—also has weight-loss benefits. Besides being a nut the coconut is also a fruit. Its juice is the water inside a young, green coconut. It's a refreshing drink, and very low in calories as the following table shows.

CALORIE COMPARISON: COCONUT WATER AND OTHER COMMON BEVERAGES

	SIZE	CALORIES	CALORIES FROM FAT
COCONUT WATER	1 cup (240 g)	46	4
CARROT JUICE	1 cup (236 g)	94	3
GRAPEFRUIT JUICE	1 cup (247 g)	96	2
ORANGE JUICE	1 cup (248 g)	112	4
PINEAPPLE JUICE, UNSWEETENED	1 cup (250 g)	140	2

Source: www.Calorie-Count.com, May 2005.

For more comparative nutritional values, see Chapter 12.

Besides being medium chain and low calorie, there are three more ways coconut oil can help you keep your weight down: good "mouth feel", increased energy expenditure, and satiety.

Coconut Oil Has Good "Mouth Feel"

Many of us do not pay much attention to the texture of the oils we use in our foods, but it is quite important. Compared to the thin vegetable oils, coconut oil has an inherent texture that gives it what is described as a good mouth feel. This basically means that when used in a recipe, 2% to 3% *less* coconut oil is needed to produce the same result as the thinner vegetable oils.

Recent Studies: Coconut Oil Increases Energy Expenditure and Satiety

There are more characteristics of coconut oil that help promote weight loss, and these were presented in a recent review of several studies comparing the effects of the long-chain to those of the medium-chain oils (8).

Studies of fats and oils often start with animals, then move on to humans. From animal studies we already know that medium-chain oils arrive faster at the liver and are more readily oxidized. Therefore, we would expect animals given diets with medium-chain oils over several months to have lower body weight and fewer fat deposits than those given long-chain oil diets. And in fact, the medium-chain oil-fed animals consistently confirmed this expectation (9-11). Newer studies have added other dimensions to coconut oil's weight-loss-promoting capabilities; these deal with energy expenditure and satiety.

Increased Energy Expenditure. There recently have been several randomized studies of the energy expenditure of human volunteers given meals composed of short-chain or long-chain oils. This energy expenditure is simply the amount of energy you use up from the food you eat. In this process, you use up oxygen and give off heat, much like the heat on the hood of your car as you drive it (12-15). The oxygen used up and the heat produced are monitored by having the volunteers stay in chambers with special gadgets that summarize their energy expenditure while on these test meals (16-17).

In one such study, the amount of energy expenditure was measured just before, and then six hours after, eating a single test meal. This showed that the energy expenditure was greater after consuming medium-chain oils than after consuming long-chain oils. *The difference measured was 48% greater in lean individuals, and 65% greater in the obese.*

In experiments studying the effect of several meals made up of medium-chain and long-chain oils, the results were consistent with the studies on single meals: as the medium-chain oils were increased, the energy expenditure was likewise increased. As the medium-chain oils were decreased, this amount likewise decreased (18).

In summary: compared to long chains, the coconut's medium chains break down to fewer ATP subunits. And compared to the long-chain energy units, using up the coconut's medium-chain units in your daily activities produces more heat. That's like a dry log that burns hotter and faster than a wet one.

Satiety . . . or How Quickly We Feel Full. In addition to looking at the amount of energy used up, the authors of the review also looked at the effect of medium-chain and long-chain fatty acids in both animals and humans on fat deposition and on satiety. How full animals felt was apparently easier to measure; those that were fed medium-chain oils in the laboratory ate less food and their body fat mass was likewise lower (19-21). In humans, the effects of medium-chain oils on satiety had variable results, although humans appeared to eat less food when long-chain oils were replaced with medium-chain oils in the diet (22-23).

Based on all the studies that they analyzed, the authors of the review paper concluded with a suggestion that—because of the greater energy expenditure, the faster onset of satiety with lowered food intake, and the resulting lower body mass—*"replacing dietary LCT (the long-chain oils) with MCT (medium-chain) could facilitate weight maintenance in humans"* (8: p. 322).

Selection Guide: A List of Long-chain Oils and Medium-chain Oils

Quite simply then, to maintain your weight or even shed weight you have accumulated over the years, try replacing the long-chain oils in your current diet with the medium-chain oils of the coconut. Coconut oil's medium-chain fatty acids are absorbed faster, provide fewer calories, produce more heat, can make you feel full faster than long-chain oils. How to get started? Use this list as a guide:

	Medium chain	Long chain
VEGETABLE AND SEED OILS	COCONUT OIL 65%	Canola and rapeseed, corn, cottonseed, olive, sesame, safflower, soybean, sunflower . . . All long chain
ANIMAL FATS		Beef, fish, all animal fats . . .
DAIRY		All long chain except butter, which has a minor percentage of short- and medium-chain oils

In a Nutshell

These five features (medium chain, low calories, good "mouth feel", increased energy expenditure, and satiety) may involve small differences, but with every meal you eat—whether fried, served with sauce or gravy, or buttered—these add up over time! Think of just one meal of a quarter-pound burger with cheese, French fries, butter on bread, and ice cream for dessert. The numbers add up pretty fast.

Unlike the long-chain fatty acids of vegetable, animal, dairy, even fish, about *two-thirds* of the coconut's fatty acids are medium chains, which means they . . .

> - are smaller, easier to digest and absorb, avoid the general circulation, get to the liver faster, are used for quick energy, and *contribute less to body fat;*
> - break down into fewer energy units and contain *slightly fewer calories* to burn per gram;
> - cause more energy expenditure while being used up for an activity;

Rx: Coconuts! (The Perfect Health Nut)

- possess a better mouth feel, so slightly less is needed in recipes;
- are more textured, so again, *less is used* in recipes; and
- give a greater feeling of fullness, faster.

Compared to other food oils, coconut oil—predominantly made up of medium-chain fatty acids—is the best oil to help you stay lean because it has fewer calories, burns energy faster than other oils, and satisfies your hunger while you eat less.

And this is only Chapter 1!

There are so many more benefits to the coconut to come. It is cholesterol free and good for your heart, it may increase your protection against cancer, it can help you fight infections and it even helps you look great.

Chapter Two

The Skinny on Saturated Fat . . . It Can Be *Good* For You

In This Chapter

- ☑ Saturated Simply Means "Single Bond" . . . Which Means "More Stable"
- ☑ Unsaturated Oils Need Chemical Processing, Resulting in Trans Fats
- ☑ The Many Health Dangers of Trans Fats
- ☑ Coconut Oil Is Trans Fat Free
- ☑ The Saturation of Coconut Oil Helps Save Your Natural Antioxidants
- ☑ In a Nutshell

Myths about food are common. That chocolates give you acne is one example. About a century and a half ago, another food myth that was deeply ingrained as common fact was that tomatoes are . . . evil! As odd as it sounds, this tomato myth was not simple rural folklore. In part, it had *medical* roots. German scientists thought that the tomato was related to the deadly mandrake, which can be used to make drugs that induce black magic; they cause people to hallucinate. The British also believed the tomato was poisonous. This myth is particularly surprising because of the tomato's long history as an excellent food, used widely by the Aztecs, Italians, French, and Chinese (it was through seafarers traveling in clipper ships to and from China that we were introduced to Chinese *kwan jup,* our modern ketchup).

Fortunately, in 1809, Thomas Jefferson, an open-minded botanist, researcher, skeptic, and by then an ex-President, began planting tomatoes on his farm at Monticello. Subsequently, he became an advocate for the tomato as food (24). And thank goodness! To this day, scientists continue to discover the health-giving carotenoids, lutein, and other antioxidants present in tomatoes.

The food myth we will be debunking here is that saturated oil—saturated coconut oil in particular—is bad for you. This idea started around the 1930s, gained ground in the 1960s, blossomed in the 1980s, and now seems as ingrained as the tomato-is-evil myth once was. Just as in the case of the tomato, the idea that saturated oils are bad for you does have a medical basis. However, as this chapter will show, new studies are challenging that simplistic conclusion.

Here, I will present scientific facts that prove that saturated oils can actually be *good* for you—not just better because of the need for less processing, but actually good for your heart and the overall health of your cells.

Saturated Simply Means "Single Bond" . . . Which Means "More Stable"

For many of us, the only thing we really "know" about saturated fat is that it is *bad*. Most of us can't explain

Rx: Coconuts! (The Perfect Health Nut)

why this is supposed to be the case, but can probably say that we heard or read it somewhere. This is, however, the kind of half-information that can lead to missing out on tremendous health benefits. For example, it is precisely because coconut oil is saturated that it is stable, remains fresh without being subjected to chemical manipulation, and is trans fat free . . . and these are all very, very *good* things indeed!

In order to understand why this is so, we again need to get a better understanding of what such terms as *saturated, unsaturated, polyunsaturated* and the like really mean. Understanding, after all, is always better than parroting the hottest buzzword and blindly following the latest trend. Armed with this new understanding, you can make a truly intelligent choice about which fats and oils to include in your diet.

Before we begin, I should let you know that some of the explanations that follow discuss a little bit of the basic chemistry of fats and oils. Many of my friends, patients and colleagues have appreciated this snippet of chemical knowledge, saying that it helped them finally understand what was what with all the oils and fats we've been told to eat or avoid, and why. If you're not interested in learning about the details of the chemistry of these fats and oils, skip ahead to the *In a Nutshell* section at the end of this chapter. If you're curious or if you're tired of the see-saw commands to "eat . . . no, avoid . . . no, eat" this or that type of fat, or if you'd like to sound smarter than the average bear at your next cocktail party . . . read on! I assure you, it's really rather painless and the knowledge you gain will help you make the smartest decisions about the foods you choose for yourself and your family.

A quick refresher from Chapter 1: all fats and oils have fatty acids. Remember that these fatty acids are made up of carbon chains and the length of these chains is what determines whether they are short-chain, medium-chain, or long-chain fatty acids. The length of the chains is important—as we saw when we followed the drop of long-chain steak oil through your system, versus the drop of medium-chain coconut oil.

Equally important are the *bonds* in the carbon chains that make up the fatty acids. The carbon chains of the fatty acids are basically a series of carbon atoms (C) connected to each other by bonds. Bonds are the special links between two adjacent carbon atoms. In addition to the *length* of the chain (which we discussed in Chapter 1), the particular *bonds* of the chain are important as well; they determine other aspects of an oil's character and behavior.

SATURATED VERSUS UNSATURATED: 101

Single Bond, Saturated, Solid (25). In order to discuss these bonds, I will need to occasionally refer to chemical signatures. Don't be alarmed; a chemical signature is merely a simple way of illustrating the makeup of a chemical by using letters and symbols (it's almost like a picture). The following chemical signature, for example, has single bonds, shown as single dashes between the Cs:

$$C-C-C-C$$

Each of these carbon atoms (the Cs) has a center and an external shell. The internal proton of the carbon atom (C) is positively charged and it is balanced at the carbon atom's outer shell by a pair of negatively charged electrons. In the chemical signature above, each of the two adjacent carbons bonds with the other's pair of electrons in a one-to-one relationship. The sharing is complete, which means that there are no excess electrons. This is important; as you will see later on, electrons that aren't paired or completely shared tend to be unstable.

The chemical signature shown above illustrates a chemical bonding that is called a single bond. Fatty acids that have only single bonds in their carbon chain are called *saturated*. And because the bonds are single and the relationship is one-to-one, this chain is more solid, stronger, and more stable. In this way, the bonds between carbon atoms act very much like real-life bonds between people. For instance, a happy couple is more likely to have a stable relationship if they are bonded to just each other, without any extra attachments. Their bonding is just like that of the stable bond between the carbon atoms of saturated fatty acids.

These single bonds give saturated oil a straight physical shape, so that the molecules are able to crowd close, to cozy in, to readily pack together into crystals. This way, they are able to assume the physical structure of a dense fat. This may instinctively sound "bad," but as you'll see later on, this physical structure is one of the most important things that makes saturated fats healthy.

Double Bond, Unsaturated, Unstable. Just like in real life, potential trouble looms when, instead of the single bond, a double bond is established. In a double bond, shown as C=C, two pairs of electrons (=) instead of one are attached to the two adjacent carbon molecules. This bond is unstable because the second electron in each pair is free. This free electron is up for grabs by any other free atom around; i.e., it is available to make reactions occur. Fatty acids with double bonds are *unsaturated*.

Unsaturated oils are further categorized as *polyunsaturated* and *monounsaturated*. Slightly unsaturated oils with just one double bond are called *monounsaturated oils* (MUFAs). Unsaturated oils with more than one double bond are *polyunsaturated oils* (PUFAs).

At the site of the double bonds of the unsaturated oils, the carbon chains tend to become kinked or curved inward into a U-like formation, making it physically difficult for the molecules to crowd together. For these unsaturated oils, crystals do not form (or form poorly), and the product remains in the form of a liquid oil, such as olive or soybean oil. Again, this is different from the saturated oils: the fatty acids of saturated oils have straight single bonds that do crystallize and that eventually create a dense fat, such as coconut oil.

Summary. The saturates are fats and oils with single bonds, they are straight, and can have the physical form of a solid fat. The unsaturates are fats and oils with double bonds, they are curved or kinked at their double bonds, and assume the physical form of a liquid oil. Simply put:

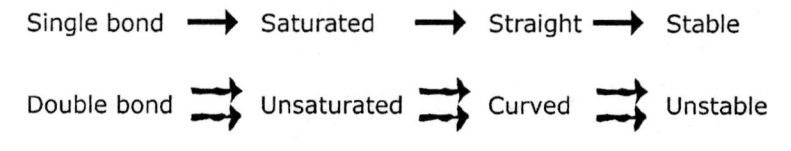

Single bond ⟶ Saturated ⟶ Straight ⟶ Stable

Double bond ⇉ Unsaturated ⇉ Curved ⇉ Unstable

Unsaturated Oils Need More Chemical Processing, Resulting in Trans Fats

Freshness, and the preservation of that freshness, is markedly different in saturated than in unsaturated oils. Coconut oil and palm kernel oil are the two major natural oils that are made up predominantly of saturated fatty acids. The presence of these saturated fatty acids in coconut oil makes it very stable even when subjected to any number of chemically destabilizing situations. For example, when subjected to light during storage or to high heat during frying, the stable single bonds of the saturated coconut oil help prevent chemical breakdown reactions from occurring.

Unlike in the saturated fatty acid, the double bonds of the PUFA and the MUFA oils make them break down, so they easily become rancid. To extend their life on store shelves and in your cupboard after you have opened the bottle, unsaturated vegetable oils invariably need to undergo a process called *partial hydrogenation*. This is a very demanding chemical and physical process. High heat and equally high pressure are used, plus, a chemical catalyst using a metal is added to force the reaction to take place.

PARTIAL HYDROGENATION: 101

Artificial Stability, Artificial Fats. To prevent unsaturated oils from easily becoming rancid, they are put through a process called partial hydrogenation.

You are now familiar with carbon atoms and how they bond with each other to make a chain. Besides the sites of bonding with the adjacent carbon, there are three more bonding sites on the carbon's shell. In fat and oil compounds, these bonds are occupied by hydrogens, another kind of atom, represented as (H). The molecule of an oil with hydrogen atoms attached, in part, appears like this:

$$\begin{array}{c} \text{H} \quad \text{H} \quad \text{H} \\ | \quad\ \ | \quad\ \ | \\ \text{H}-\text{C}-\text{C}-\text{C} \\ | \quad\ \ | \quad\ \ | \\ \text{H} \quad \text{H} \quad \text{H} \end{array}$$

Partial hydrogenation affects these hydrogen atoms. This process subjects the unsaturated oils you use in your food to the high heat of a furnace, to force the entry of more hydrogen atoms into the oil's fatty acid chains. This physical manipulation is meant to change the natural molecular structure of the double bonds of the fatty acids. Some double bonds become single bonds to imitate the single bonds of the saturated coconut oil; many others remain as double bonds, but these have now become very unnatural.

Remember that the natural shape of the unsaturated fatty acid at its double bonds is a curve or a kink. At this curve or kink, the hydrogen atoms both stay on the same side, whether above or below the carbons with the double bonds. This is called the *cis* position of the hydrogen to its carbon atoms. Because of the chemical and physical process of partial hydrogenation, the hydrogen atoms rearrange themselves, moving from their natural position (cis) to a new position wherein they are located across from each other. This position is called *trans*.

To repeat, with partial hydrogenation, most of the double bonds of the unsaturated oils remain, but their hydrogen atoms change position and become trans, or across from each other. To emphasize: *cis* is the natural position and *trans* is the abnormal position.

The natural cis position is like a plié or pirouette—the ballerina's body bends and has kinks and curves. For the trans position, imagine an arabesque or a ballet split that allows the body—or in this case, the carbon chain—to become straighter, to unbend and remove the kinks and curves. The two hydrogen atoms are at opposite sides, one above and one below the carbon atom. The straightening of the fatty acids allows the hydrogen atoms to come closer together. Below is a simple diagram of the positions of H and C atoms at the double bonds of unsaturated fats.

Cis

Sample of a moleculue with a U-shape

Trans

Sample of a Straightened (ballet-split) shape

Partial hydrogenation transforms the PUFA oil into a more solid fat, like the natural saturated fat of the coconut. This straightening is an artificial creation to make the vegetable oil into a shortening or margarine. This artificial straightening is necessary for two commercial purposes. First, the liquid is converted to a lardlike texture (this imitates a natural transformation of the coconut oil, which we will discuss further on). Second, and equally important, like the naturally single-bond saturated coconut oil, the margarine is now more stable and can be kept at room temperature. It can therefore safely stay on the supermarket shelf or in your cupboard long after you have opened the container.

However, partial hydrogenation comes at a high cost. This process produces trans fats, a synthetic or artificial dietary fat that our body cannot naturally process. Partial hydrogenation changes as much as 40%, 50%, or more of the oil's natural fatty acids into trans-fatty acids. And trans fats have been conclusively shown to be the source of many health problems.

The Many Health Dangers of Trans Fats

First and foremost, trans fats have a negative effect on the circulatory system. Trans fats lower the levels of "good" HDL cholesterol in your blood while raising the levels of LDL, the "bad" cholesterol. Low levels of HDL (which *should be high*) and high LDL (which *should be low*) are well established to be associated with blood vessel diseases that can result in a heart attack or a stroke (26-33).

The evidence against trans fats is so strong that in July 2003, the FDA mandated the disclosure of the amount of trans fats on the labels of all processed foods (34). Forewarned in this manner about the level of trans fats that a product contains, we can make our choices accordingly. Do keep in mind, however, that any inclusion of trans fats in your diet increases your risk of various health problems.

More Dangers of Trans Fats. Besides the negative effects on the heart, there are other health problems reported from including trans fats in the diet. Since the 1970s, researchers have warned us that trans fats in the diet negatively affect our health (35-38), starting with the expectant mother (39) and the milk she produces (40), and then affecting her baby (41) and her older children (42). Trans fats are also reported to affect general metabolism, making people more prone to obesity, diabetes, and even to immunologic problems and asthma (43).

Trans Fats Cannot Be Processed by the Body. This is an important reason for the negative effects of trans fats. The human body does not have the protein enzymes needed to break down these artificial fats. In a sense, trans fats are like all the plastic bottles and bags and debris we have a hard time dealing with in the environment. This debris is often indestructible, clogging up drainage and sewer systems. Just think what the debris from the synthetic oils—trans fat—does inside your body.

Which Foods Have Trans fats? Despite these health risks, trans fats make up as much as 40% or more of partially hydrogenated vegetable oils, which are a staple of the worldwide food industry. These oils are widely consumed in restaurant food and in bakery goods, as well as in chips and other increasingly popular snack foods.

Some manufacturers are now starting to use unhydrogenated oils that are subjected to a different chemical process that eliminates the odors of the oils (read the ingredient statements). However, this process is still a refinement of the PUFA oils and may produce some trans fats.

Cookies, crackers, donuts, cakes, bread frostings, puddings, snack chips, imitation cheese, croissants—all use margarine, artificial butters, or other commercial shortenings made with partially hydrogenated oils. Trans fats are also accepted as ingredients in home cooking, baking, and frying (44, 45).

Coconut Oil Is Trans Fat Free

One more delicious fact about the coconut: coconut oil has no trans fats. Because coconut oil is saturated and more stable, it does not need to go through partial hydrogenation. In fact, what partial hydrogenation tries to get the unsaturated oils to copy, coconut oil does naturally. Coconut oil has an inherent yin and yang trait: it effortlessly switches from one physical form to another through the simple rise or fall of a room's temperature. In the tropics, and at room temperatures at or above 25°C (76°F), coconut oil is a clear liquid. In cooler climates or in the refrigerator, the oil readily becomes a pristine white butter. The texture of this coco butter is ideal for use in baking.

Summary. The switch from oil to fat to oil is a physical transformation quite natural to the saturated coconut. At lower than tropical room temperature, or in the refrigerator, coconut oil naturally transforms into a rich butter, which you can use in baking.

PUFA oils tend to always remain in the liquid oil state. They need partial hydrogenation to transform their oily texture to that of butter, lard, or margarine. This partial hydrogenation of PUFA oils creates 40% or more trans fats. These artificial fats are very bad for your heart and general health.

Because coconut oil does not need to go through partial hydrogenation, it has no trans fats.

The Saturation of Coconut Oil Helps Save Your Natural Antioxidants

When you eat coconuts, the saturation of coconut oil becomes an even greater asset for your health, not the threat of popular myth. Keep in mind the bonds analogy and their effect on couples. As between healthy couples, saturated oils have single bonds and make a stable partnership, while unsaturated oils have double bonds that create a potentially disturbed one. This characteristic of single versus double bonds has an important impact on your general health, as we will see below.

How Saturated Fats Help Prevent Harmful Oxidation. Besides providing energy, the fatty acids you eat are used to make the membranes of your cells. The fat content of your cell membranes ought to be balanced at about 50% to 60% unsaturated fatty acids, 40% to 50% saturated fats, and some cholesterol.

To illustrate the importance of saturated fats for your cell membranes, let's look at what can happen when this ratio is made unbalanced . . . as when one pursues a *low-fat* diet. At one point or another you, like many of us, may have been told to have a healthier diet by eating food containing less fat and less cholesterol. When you adopt these low-fat diets, you usually eat fewer animal fats and dairy products, reducing your saturated fat intake to about 20%. However, the unsaturated fatty acids in your diet then go up to as much as 80% because the foods you substitute for saturated fats are instead high in PUFA oils! When this happens, the resulting fat content ratio of your cells' membranes is not very healthy, as I discuss in the next chapter. As you will later see in that discussion, it is often healthier to be more selective about the *kinds* of fat you eat.

For now, let us return to our example of the low-fat diet and the discussion of how the decrease of saturated fatty acids and tremendous increase in unsaturated fatty acids becomes reflected in the fats-to-oils ratio of your cell membranes. While you are on this low-fat diet, and as the number of unsaturated fatty acids in your cell membranes increases, your cell membranes become vulnerable to oxidative attack.

Oxidation in the body can be a normal process such as the one we discussed in Chapter 1—beta-oxidation. Recognize first that the carbon chains of fats and oils have even numbers: C_6, C_8, C_{10}, etc., up to C_{24}. In this type of oxidation, oxygen and chemicals called enzymes, starting at one end of the carbon chain, break off two carbon units. These units then become your body's gasoline, ATP energy molecules. This oxidation not only is normal but also is beneficial, as it creates the energy you need.

Harmful oxidation is induced by highly unstable chemicals called free radicals. These free radicals are a regular by-product of the body's daily machinery needs, much like the exhaust from your car as it burns up gasoline. Free radicals also can be produced through aging, disease states, physical and emotional stress, sun exposure, or strenuous exercise.

Some free radicals actually can be useful—such as when certain cells in your body make them to help your immune system fight off invading bacteria and viruses. More frequently, however, these free radicals are like harmful guerrilla fighters that constantly look for and attack your weak spots. One such weak spot is at the unsaturated PUFA oils of your cell membranes. The atoms in all cells have a positively charged center and a pair of negatively charged electrons circling around it. At the double bonds of PUFA oil atoms, one electron is attached to a carbon while one orbiting electron is unpaired. This unpaired electron is like a huge magnet for a free radical.

Because circulating free radicals have single electrons, they frantically seek electrons to pair with their unpaired electrons. They may even steal electrons from normal molecules that already have paired electrons. The unpaired electrons from the side chains made up of the PUFA oils are therefore easy prey for these free radicals.

To Recap. In fatty acids, the bonds between carbon atoms contain two electrons each. In the single bonds of saturated oils, both electrons have a one-to-one relationship and are stable. In the double bonds of PUFA oils, one of the two electrons is unpaired. It is this unpaired electron that is unstable, that becomes a weak spot in your cells' membranes, and that can be targeted by harmful free radicals. Remember that in our example of a low-fat diet, your food intake tends to skew the ratio of the fats that make up your cell membranes so that they have significantly more PUFA fatty acids than saturated fatty acids. Here, you see that the more PUFA oils you intake >> the more PUFA fatty acids make up your cells' membranes >> the more double bonds are present >> the more unpaired electrons >> the more weak spots and easy targets for free radicals.

Prevent Harmful Oxidation: Take Antioxidant Action Just by Substituting Coconut Oil for Less Healthy Oils in Your Diet. At the cell membrane, the harmful oxidation reaction discussed above is called peroxidation, and it produces a specific type of free radical, the peroxyl radical. Peroxyl radicals are amazingly destructive. Created from a fatty acid, they can in turn attack another fatty acid, setting off a domino-like effect of chemical chain reactions among more fatty acids and other molecules.

While the PUFA fatty acids are especially vulnerable, even saturated fatty acids, proteins, sugars, and hormones in our bodies, and the DNA in our cells, can be attacked by these peroxyl and other free radicals.

Your body is equipped with natural antioxidants and repair mechanisms to take care of the effects of these free radicals and harmful oxidation reactions. But when free radicals form and harmful oxidation occurs in excess, it all results in some kind of destruction or disturbance of the structure and function of the cells and of the organs to which they belong. In the long run, healthy cells become less healthy; they die or they change their structure and, therefore, change the way they act. Cell oxidation is blamed, in part, for cancers and for degenerative conditions, including Alzheimer's disease and aging.

The now widespread use of antioxidants in food and as medicine is one response to the newly recognized role of oxidation in producing disease. So, taking antioxidant supplements is a good idea to take care of the free radical damage from causes you cannot avoid. Another proactive response is to *change the ratio of PUFA oils to saturated fatty acids in your cell membranes.*

What's a simple way to do this? Remember that the fatty acids of coconut oil are mostly saturated and stable. Almost all (about 92%) of the fatty acids in coconut oil are saturated, and only about 8% are unsaturated. At the cell membrane, therefore, they are not as vulnerable to peroxidation as the PUFA oils. Simply increasing your saturated fatty acid intake by using more coconut oil in your diet can minimize peroxidation at the cell membranes and can help you save your precious natural antioxidants.

In a Nutshell

Saturated oils like coconut oil, with more stable, single bonds in their carbon chains, have several health advantages.

- Coconut oil remains naturally fresh longer
- Coconut oil naturally becomes an oil in a warm room and a white butter in cooler temperatures, making it useful in baking without having to go through partial hydrogenation

- Because coconut oil needs no partial hydrogenation, it is *trans fat free* and *heart-healthy*
- Because coconut oil is a saturated fat with single bonds it is more stable and *less prone to oxidation* by free radicals that destroy cells and promote disease

Far from being negative, it is precisely coconut oil's status as a *saturated* oil that allows it to be more stable, free of chemical processes, free of trans fats, and *good* for your overall health and your heart—which we discuss in detail in Chapter 3.

Chapter Three

Coconut Oil Is Cholesterol Free, Trans Fat Free, and Heart-Healthy

In This Chapter

- ☑ Learn From History: Coconut Oil in Our Food = Fewer Heart Attacks and Strokes
- ☑ Population Studies Show People Who Eat Coconut Oil Regularly Are Heart-healthy
- ☑ Coconut Oil Is Cholesterol Free
- ☑ What Is Bad for the Heart? Too Many Hidden Carbs, Linoleic Acid, Blood Clotters and Trans Fats
- ☑ What Is Heart-Healthy? Cholesterol-free, Low-carb, Trans Fat-free, Low-linoleic Acid Coconut Oil
- ☑ In a Nutshell

We've talked about how the coconut can help make, and keep, you slimmer. And we've seen how many of the health benefits of the coconut are actually due to its being a saturated fat—benefits such as being more stable, not needing chemical processing, being trans fat free and helping to save your natural anti oxidants. Now let's look at how the coconut's bevy of benefits can be applied to your heart. Heart disease is, after all, one of the biggest health concerns worldwide. What we will see in this chapter is that the coconut's use is linked to a heart-healthier period in our history, that populations that tend to eat coconut oil and coconut products have fewer heart-related diseases, and that coconut oil has several unique characteristics that make it excellent for your heart.

Learn From History: Coconut Oil in Our Food = Fewer Heart Attacks and Strokes

For good reason, George Santayana's edict, "Those who do not learn from history are doomed to repeat it" is one of the world's most recognized and repeated. The man and his message are appropriate here as well. Born in Madrid in 1863, this famous American philosopher, poet, and humanist lived through the years when coconuts were a regular part of the American diet. He died in 1952 from unknown causes, though very likely from the complications of old age, having reached his ninetieth year (46).

About one hundred years ago, around the turn of the twentieth century, the American diet was *high in fat*. Commercial shortenings advertised to housewives contained butter, lard, or tallow from animal fats. After winning the Spanish-American war in 1898, the United States acquired the Philippines as part of the Treaty of Paris. Shortly thereafter, increasing amounts of coconut oil became available from its new colony. This was used in soaps, as a cheap replacement for the animal fats in butter, for margarine, and even for gunpowder.

By the first quarter of the twentieth century, coconut oil accounted for as much as 90% of commercially sold margarine (7). Yet, during this same period, cardiovascular disease was virtually nonexistent!

Dr. Paul Dudley White, the famous heart surgeon, loved to tell this story, repeated by many: in 1912, when he returned from subspecialty training in cardiology in Europe, he was told that there was no need for his newly acquired ECG heart-monitoring machine. At that time, there was hardly any heart disease in the United States. It was not until several years later that he saw his first case of a heart attack.

Also around the turn of the century, the domestic edible oil industry was rapidly developing. These oils are liquid, and because they are PUFA they easily become rancid. In 1910, Partial Hydrogenation ushered in a new era for vegetable oils because, as we have seen, this process makes the oils acceptably lardlike and more stable. However, as we have also seen, this came at a price: the process introduced a high amount of trans fats, the dire effects of which were only realized by Dr. Mary Enig and other researchers sixty years later (47).

By the 1930s, animal oil producers began to use their political clout to push coconut oil out of the market. First, there was a political campaign, supported by the vote-heavy U.S. dairy industry, which resulted in a 1934 excise tax of three cents per pound on coconut oil (48, 49). Coconut oil rapidly dwindled in American diets, and just as quickly, the PUFA oils replaced it.

Then, as I mentioned briefly in the Introduction, from the 1960s through the 1980s, the vegetable oil industry undertook a major campaign to further exclude coconut oil form the American diet, and was consequently successful in bringing vegetable and seed oil to dominance. Coconut oil food consumption continued to disappear, down to about 3% of American diets (50, 51).

From the turn of the twentieth century to the time the coconut was phased out of American diets, heart attacks were rare. Here are the figures. In 1900, the United States reported only 345 deaths per 100,000 per year from coronaries. By 1960, with coconut oil down to and remaining at about 3% of American diets, the number of deaths per 100,000 went up to 522. With numerous modern surgical, medical, pharmaceutical, interventions, and with coconut oil still remaining at about 3% of

American diets, in 2001, 863,680 deaths (42% of all deaths, compared to 8% in 1900, and 18.9% in 1930) were from cardiovascular disease (52).

The lesson of history is clear : coconut oil, long gone from our diets, is not the culprit.

Population Studies Show People Who Eat Coconut Oil Regularly Are Heart-healthy

Population studies are like natural experiments using the largest possible number of subjects. A series of population studies that involved the United States, Europe, and Japan in the 1960s was taken as proof that saturated animal fat was the cause of the rapidly increasing incidence of coronary heart disease. These studies, initiated by Dr. Keys in 1957, became the linchpin in the diet-heart theory of heart disease, which prevails even today (53).

Similar population studies for the same time period, however, show that populations consuming high levels of coconut oil have low illness and death rates from coronary heart disease. Moreover, among these coconut-eating population groups, the blood cholesterol levels of those who were tested correspond with these low heart disease rates (54).

Let's start with the example of Sri Lanka, a high coconut oil-consuming country. The *Demographic Yearbook* of the United Nations is a good place to begin, looking at death rates per 100,000. In 1978, Sri Lanka reported a death rate of only 1 from heart disease due to clogged blood vessels. The 1978 death rate in countries with little coconut oil consumption varied from 16 to 187 (55). Noting the Sri Lanka experience, Drs. Mendis and Wissler became curious about the cholesterol levels in the blood of Sri Lankans (56). Sri Lankans' usual daily diet has a lot of coconut oil. The doctors therefore expected their blood cholesterol levels to be high. Instead, they found the blood cholesterol values of sixteen healthy Sri Lankan males, studied over six weeks, to be low. The good HDL cholesterol and the bad LDL cholesterol were both at desirable levels, with an LDL:HDL cholesterol ratio of *3:1*. This is a good ratio, and certainly correlates with the Sri Lankans' reported low rates of blood clots in the heart.

Still curious, the doctors continued the study. They gave these young men a diet of cow's milk powder and corn oil, to replace the coconut oil. After six weeks on this diet, the men's LDL cholesterol values fell, which is excellent, because this is the bad cholesterol. But the HDL—good cholesterol—decreased much more, bringing the LDL:HDL cholesterol ratio to about *4:1,* which is undesirable.

These values correlate with the reported increase in the numbers of coronaries among coconut-eating people who move out of the tropics to places with a diet no longer including much coconut oil (57, 58).

For many years, the American Medical Association recommended that our diet should consist of not more than a total of 30% total calories of fat, only 10% of which should be saturated fat. Dr. Prior et al. (59) studied the blood lipid chemistries of Polynesians, another group eating a lot of coconut. They studied two population groups, the Pukapukans and the Tokelauans. The Pukapukan males ate 32% of their total calories in the form of fat; females ate 39%. Of this diet, 75% was saturated fats, mostly from coconut. The Tokelauans ate even more fat—56% of total calories for both males and females—90% of which was from coconuts.

The Pukapukans, who ate as much fat as most Caucasians but from the saturated coconut, had a mean cholesterol level below 180 mg/dl. The value for the Tokelauans was slightly higher, though still acceptable, at 208 mg/dl for males and 216 mg/dl for females. Again, these coconut-eating groups were also reported to have few coronaries.

And here's more. In 1984, Drs. Florentino and Aguinaldo (60) surveyed nine regions of the Philippines, another coconut-eating country. Filipinos in general were found to eat fewer fat calories than Americans. Of their fat intake, coconut oil consumption ranged from a low of 36.7% in metropolitan Manila to a high of 62.4% in the Bicol region. Famous for their coconut-based dishes, every Bicolano's food, it seems, has some coconut milk in it. They ate 26 grams of coconut daily, compared to the average Manilan's intake of 16 grams daily.

The cholesterol level among Bicolanos was the highest among Filipinos, although the values were still below the accepted norm of 200 mg/dl. Even more significant is that among all Filipinos, these Bicolanos had the lowest number of deaths from heart disease and strokes.

How then, you may ask, does total heart disease in the Philippines compare with that in other countries? Dr. Dayrit examined Dr. Keys's paper, which compared total heart disease deaths per 100,000 deaths to the intake of fat in six countries, using the data for the years 1950-1952. Dayrit plotted a similar graph adding fifteen more countries with data also available for the years 1950-1952. To this graph he added the Philippines' heart disease rates. The Philippine data for the years 1950-1952 showed infectious diseases as the leading cause of death, with heart disease rates ranking much lower. By 1982, the country's effective infectious disease control had decreased the ranking of deaths from infections, such that heart disease rates had gone up to a higher rank, more comparable to the 1950-1952 ranking of diseases in other countries. Dr. Dayrit therefore used the Philippines' 1982 data (54).

The map is visually startling. Among the twenty-one countries, the Philippines is the only coconut oil-consuming country. If coconut oil is really bad for the heart, the Philippines should be among those with the highest death rates from heart disease. Instead, it has the *lowest* heart disease death rate, lower even than that of Japan, which previously was the lowest.

There are many other population studies coming from countries where the coconut has been used as food for centuries (61-63). Many studies come from India, the country where the ancient practice of holistic medicine called Ayurveda began and continues until today. Among those who follow this form of spiritual and medical practice, the coconut continues to be highly respected as food, as medicine, and for good health (64-67).

In addition, through the years, there have been smaller studies examining whether coconut oil raises cholesterol and as a consequence, according to the diet-heart theory, blocks blood vessels. Kintanar (68) used evidence-based standards to review 119 original coconut oil articles. Some were animal experiments, others were human clinical studies, review papers, and the population studies from the 1960s to the1980s.

Only about one-fourth (27%) of the studies showed that coconut oil raised cholesterol levels. But those studies had substantial flaws, the first being the number of participants in the studies. Whether animal or human, some of the studies had only a handful of subjects. The second flaw was the use of partially hydrogenated coconut oil—which, I should emphasize,

Rx: Coconuts! (The Perfect Health Nut)

is *not* the coconut oil we recommend; the healthy coconut oil to which this book refers is unhydrogenated virgin coconut oil.

As a quick aside, even if partially hydrogenated coconut oil were to be used it has only about 8% unsaturated fatty acids (like all coconut oil), about half of which, almost 4%, may becomes trans fats after partial hydrogenation. Compare this 4% possible trans fats to the possible 40% trans fats in unsaturated fats after partial hydrogenation. Let me reiterate, however that it is virgin coconut oil and not partially hydrogenated coconut oil that I recommend to totally avoid the possibility of trans fats.

More importantly, coconut oil only has a small amount (2%) of an essential fatty acid called linoleic acid. This is totally destroyed by the hydrogenation. Since the body needs to receive 2% to 3% of its calories from this essential fatty acid, its absence in the hydrogenated coconut oil resulted in an essential fatty acid deficiency in the test subjects.

Just as with you and me, stress increases cholesterol levels, and this also increased in the nutritionally stressed, essential fatty acid-deficient laboratory animals. These high cholesterol levels were wrongly attributed to the animals' intake of coconut oil rather than to the deficiency that developed from overlooking the addition of essential fatty acids to their diets.

Unhydrogenated coconut oil should have been used, and the essential fatty acids that coconut lacks should have been added to the diet. Some studies actually did this. As expected, when there was no essential fatty acid deficiency, the animals did not develop hardened blood vessels (69-72).

You may be asking, Does this mean that if I just use coconut oil, I, too, will become deficient in essential fatty acids? The answer is no. In any regular human diet, a little fish, nuts, or vegetables readily supplies these fatty acids. We only need about 3% to 5% of the two essential fatty acids in our diet.

The rest of the studies showed the coconut to be, at worst, neutral to the cholesterol values. In some, the good HDL cholesterol even went up (73). A recent study confirms the positive effects on HDL and LDL values and the heart, just as was demonstrated by the older coconut oil studies. One study looked at the lipid profile of human volunteers eating a solid-fat diet rich in lauric acid versus a group eating a diet rich in trans-fatty acids (74).

Conclusively, trans fats were found to have stronger adverse effects on cholesterol values than the lauric acid.

One study in 1991 (61) and two studies in 1994 (75, 76) similarly showed increased total cholesterol, a minimal decrease of LDL, and a more significant increase of HDL, resulting in a lowered LDL:HDL ratio. These values mean a favorable effect of diets high in lauric acid (abundant in coconut oil). Enig concluded: "studies that supposedly showed a hypercholesterolemic effect of coconut oil feeding, in fact, usually only showed that coconut oil was not as effective at lowering serum cholesterol as was the more unsaturated fat being compared. This appears to be in part because coconut oil does not *drive* cholesterol into the tissues as does the more polyunsaturated fats (51)".

The last part of her quote is based on the chemical analysis of the fatty concretion called *atheroma*, that blocks blood vessels, the prelude to a heart attack. The fatty acids in these atheromas are 74% unsaturated, 41% of which are PUFAs. Only 24% are saturated fatty acids—none of which are lauric acid or *myristic* acid (77).

The lesson from population studies again is clear: populations that have been eating a lot of coconut in their diet for centuries are heart-healthy.

Coconut Oil Is Cholesterol Free

Let's break this down as simply as possible:

- ➤ Cholesterol has been implicated as a cause of heart attacks and strokes.
- ➤ The *saturated* oils that come from *animals* are *cholesterol rich.*
- ➤ Coconut is a *saturated oil* that comes from a plant and is *cholesterol free.*
- ➤ Therefore, by virtue of its being *cholesterol free,* the coconut is heart-healthy.
- ➤ As for its effect on blood tests for *total cholesterol* and *LDL cholesterol* levels, used by doctors to predict heart and vessel disease, the results for the coconut are not yet clear.

Rx: Coconuts! (The Perfect Health Nut)

Why Is "Cholesterol-Free" Important? As we've done at every stage of this book thus far, let's try to get a better understanding of just how the recommendation of "cholesterol free" came about, so as to make more intelligent dietary choices. The diet-heart theory was first preached in the 1950s by the physiologist Dr. Ancel Keys (53). According to his theory, *all* saturated fats are high in cholesterol. High blood cholesterol is a major factor in coronary heart disease. Therefore, all diets high in saturated fats or cholesterol were to be avoided.

In line with this logic, doctors' orders became: Avoid all foods high in saturated fats, because they are high in cholesterol. When your serum cholesterol goes up, you can become the next victim of a stroke or a coronary (78, 79).

This theory was so widely accepted that by 1992 the USDA built this advice into its Food Guide Pyramid program (80). Carbohydrates are cholesterol free; therefore, they were to form the recommended bulk of everyone's diet, making up the wide base of that pyramid. Proteins from chicken, fish, and meat came higher up in the pyramid, and were to be eaten less.

Some of you may have heard the myth that "coconut oil is bad for your heart," and this is in part how the myth got started. Just like other aspects in the Food Guide Pyramid, that coconut oil is bad for the heart has been wrong all along. Just consider this . . . all fats and oils, in particular the saturates, were considered bad because of their cholesterol content and were to be minimized, which placed them at the small apex of the food pyramid. What happened to coconut oil in this food pyramid? The coconut is a saturated oil, but as I have already noted, it comes from a plant and is therefore cholesterol free. Since the pyramid facts do not point out this distinction, people interpret coconut oil as belonging to the group of saturated, cholesterol-rich fats from dairy products and animal fats. The oils considered acceptable for consumption are the other, also cholesterol-free, plant oils. These are the monounsaturated (remember, the MUFA) and the polyunsaturated (PUFA) oils from soybeans, corn, cottonseed, rapeseed-canola, olive, and other seed oils.

After twelve years of heavy publicizing and promotion of the food pyramid, with coconut oil still composing only about 3% of the American diet, the diseases it was supposed to eliminate—heart disease, hypertension, heart attacks, and strokes—are still common. Worse, more diet-related problems such as diabetes and obesity have become pandemic (81, 82).

Throughout these chapters, I have often emphasized that coconuts are cholesterol free. I shall leave to others the subject of the role of cholesterol from the saturated animal fats and dairy products in your diet on the blood vessels of your body (83, 84). Cholesterol does have a role, and there are blood tests that allow your doctor to recognize high cholesterol fractions that may be bad, so she or he can warn you about what you should do and the medicines that you need to take.

However, just as important are the many other factors, not necessarily related to diet, that can predispose you to have a stroke or a heart attack. For these many years, these have not been given the same attention that has been given to cholesterol. Fortunately, this is changing, though not fast enough. Dr. Conrado Dayrit calls the major underlying causes the Four I's of human disease: inheritance (genetic), infection, immunology, and inflammation (85). Obviously, the coconut cannot help you in the area of how your genes were selected. But infection, immunology, and inflammation all respond favorably to proper diet.

The first lesson here is that coconut oil should not have been lumped together with other cholesterol-rich saturated animal and dairy fats, because it is in fact cholesterol free. The second lesson is that, even as coconut oil consumption in the United States has remained at lower-than-ever rates, heart disease (which the food pyramid was supposed to help curtail) has remained common, and new diet-related diseases have soared. The third lesson is that although cholesterol does play an important part, it *alone* is not the only thing to consider when pursuing a heart-healthy diet.

What Is Bad for the Heart? Too Many Hidden Carbs, Linoleic Acid, Blood Clotters and Trans Fats.

I f the coconut is heart-innocent which items in the diet are heart-guilty? The culprits are not just the saturated fats, after all, and not cholesterol levels per se. Instead, these diet-related problems can be summarized in two words: *too much* (52).

Rx: Coconuts! (The Perfect Health Nut)

- *Too much* refined carbohydrate.
- *Too much* polyunsaturated oil (PUFA).
- *Too much* linoleic acid
- *Too much* potential blood clotters
- *Too much* of the trans fat from partially hydrogenated PUFA fats.

Let us look, one by one, at the things that became *Too Much* in our diets, even unbeknownst to us.

Too Much and Hidden: Refined Carbohydrates. We ate more carbohydrates because we were told to do so. But we ate many more of them than we knew, hidden, in the form of refined carbohydrates designed to replace the taste of the lost saturated fats. So that you can relate to what *refined* means, think of white sugar versus brown sugar, white polished rice versus brown rice, or white flour versus brown flour. Refined, in other words, means processed. In each of these examples, the processed product was perceived to be of better quality, more visually appealing. Unfortunately, the physical and chemical processes used to "refine" the product remove many of its natural vitamins, minerals, and nutrients.

Am I going to make your diet a simple matter of food color? Tempting, but no. I merely want to point out that more of these refined carbohydrates have been added to the "low-fat," less tasty foods you were told to eat more of—in order to replace the natural taste of fat and because of the carbohydrates' great mouth feel. These carbohydrates are hidden in the foods that carry them, like the fine print that we tend to ignore in fifty-page contracts. Also hidden is the fact that these carbohydrate products are processed using margarine and shortening, which are already high in trans fats. Hidden refined carbs, made with stuff that's high in trans fats: a double whammy.

Unaware, we drank skim milk by the gallon. Through the years, skim milk became tastier, and we never asked why. Now we know that the sweeter flavor of skim milk results from the addition of refined sugar or chemicals with names that do not *say* sugar but that are—yes—refined carbohydrates. The flavor of most, if not all, other low-fat foods also comes from these refined carbohydrates.

63

Look at the Carbohydrates section of the Nutrition Facts on the back of the carton of the fat-free Half-and-Half or the Light Cream in your refrigerator. Chances are they have 1 to 3 grams of those hidden carbohydrates per serving. Then look at heavy cream. It has *zero!* Look also at 1 cup of whole milk. It has *11 grams* of carbohydrates, whereas skim milk has *12 grams* (53). One-third cup of nonfat dry milk—which, after water is added, is equivalent to 1 cup of whole milk—also has *12 grams* (53).

We complied with the mantra "eat low fat," then guiltlessly stuffed ourselves with more pasta and more bread, both made up of refined flour, or we ate equally refined rice. Too late, we realized that these refined carbohydrates do not quickly provide the feeling of fullness that proteins, fats, and oils do. The food servings on our plates steadily grew and our total calorie intake, not to mention our carbohydrate intake, went up to ever-higher levels.

These carbohydrates are rapidly used for the energy you need to keep your body's machine going—at rest, at work, and at play. Just like the processing of fats, carbohydrates also are processed into the small chemical subunits your cells can use. Unlike coconut oil, these subunits demand the use of insulin. Produced by an organ called the pancreas, insulin circulates until it reaches points of contact on your cells called insulin receptors.

When you consume too many carbohydrates, the insulin receptors become overused. Eventually, these receptors are damaged. When this happens, your cells become resistant to the insulin. As you continue to eat more carbohydrates, this insulin resistance in turn leads to a need for more insulin, the body produces more of the hormone, damages more receptors, and the vicious cycle of hyperinsulinism begins.

When this happens, you are headed toward what doctors label Syndrome X. This state of being may sound exotic, but you already know a fair number of people who have this, and they are not very happy. They are grossly obese and have pendulous stomachs. In their blood, they have low HDLs—the good cholesterol—and high LDLs—the bad cholesterol. And they end up being treated for diabetes, high blood pressure, and/or heart disease.

Too Much and Hidden: Polyunsaturated Fats (PUFAs). Just as we eat too much of the carbohydrates, we also tend to eat too much of the PUFA fats (7). They are good oils, but we consume too many of them: at breakfast, at lunch, at dinner, and in snack foods. The problem with eating too much PUFA oil is that you get too much of a potent fatty acid called *linoleic acid*. At least 70% of the PUFA fat you eat is linoleic acid.

This linoleic fatty acid is actually one of the two essential fatty acids you need to include in your diet, because your body does not produce them. But as I mentioned earlier, you only need about 2% to 3% linoleic acid in the fats that you eat. The importance of linoleic acid is that your body uses it as a startup fatty acid for the making of more specialized fatty acids, including a special one called arachidonic acid. What's with that acid? Arachidonic acid is one component of a special kind of fat called *phospholipids*. Phospholipids have two basic functions: to help form the membranes, or walls, around and in each cell in your body, and to help produce a biologically active group of substances called *prostaglandins*. Hang in there; I'll stop with the big names after this.

Let me just explain that prostaglandins are tissue hormones, which are vital to the regulation of the function of your cells. There are several of these tissue hormones. Some have similar functions, while others directly oppose these functions. This maintains the checks and balances to fine-tune the function of your cells. Here is where too much of a good thing can be bad. When arachidonic acid is present in abundance, as it may well be from the excess linoleic acid of the PUFA oils you consume, your body may produce more of the prostaglandin compounds that favor the clotting of blood, or the constriction of blood vessels. These can pave the way for a heart attack or a stroke.

So, again, too much of the linoleic and arachidonic acids can be bad for your blood vessels and your heart. Coconut oil has very little of these pro-clotting linoleic and arachidonic acids (86). The seed and vegetable oils have plenty, and they are often hidden in the processed foods that we frequently eat (43-45).

Too Much and Hidden: Trans Fats. Most importantly—a point previously discussed but worthy of singling out and repeating—too many of the trans fats from the partial hydrogenation of PUFA oils are *really* bad for the heart. Likewise worthy of repetition, coconut oil is trans fat free.

What Is Heart-healthy? Cholesterol-free, Low-carb, Low-linoleic Acid, Trans Fat-free Coconut Oil

Coconut oil is heart-healthy.

- Compared to other edible nuts coconuts are *naturally lower in carbohydrates.*
- Coconut oil is *cholesterol free.*
- Coconut oil is very *low in linoleic acid,* which produces pro-clotting prostaglandin chemicals.
- Coconut oil is *trans fat free.*

So is it just the oil of the coconut that's good for the heart? Hardly. Besides being a nut, the coconut is a vegetable and a fruit and can yield a heart-healthy juice (commonly called coconut water) and meat or flesh (similar to that of any other vegetable or fruit). The wonder of it all is that both the water and the meat taste great yet are naturally low in carbohydrates. The following two tables illustrate the coconut's lower carbohydrate count when compared to other fruit juices, vegetable juices, and edible nuts.

NATURAL CARBOHYDRATE COMPARISON: COCONUT WATER AND OTHER FRUIT OR VEGETABLE JUICES

NUTS	SIZE	TOTAL CARBOHYDRATE	DIET FIBER	SUGAR
COCONUT WATER	1 cup (240 g)	8.9 g	2.6 g	6.3 g
CARROT JUICE	1 cup (236 g)	22 g	2 g	9 g
GRAPEFRUIT JUICE	1 cup (247 g)	22.7		
ORANGE JUICE	1 cup (248 g)	25.8	0.5 g	20.9 g
APPLE JUICE	1 cup (246 g)	29 g	0.2 g	
PINEAPPLE JUICE	1 cup (250 g)	34.5 g	0.5 g	34 g

Source: www.Calorie-Count.com, May 2005.

Rx: Coconuts! (The Perfect Health Nut)

NATURAL CARBOHYDRATE COMPARISON: COCONUT MEAT AND OTHER EDIBLE NUTS

	SIZE	TOTAL CARBOHYDRATE	DIET FIBER	SUGAR
COCONUT MEAT (RAW)	1 cup (100 g)	15.23 g	9 g	6.23 g
COCONUT MEAT (SHREDDED)	1 cup (80 g)	12.2 g	7.2 g	5 g
WALNUTS	1 cup shelled 50 halves (100 g)	14 g	7 g	3 g
PECANS	1 cup (119 g)	16.5 g	11.4 g	4.7 g
PEANUTS, DRY-ROASTED, NO SALT	1 cup (146 g)	31 g	12 g	6 g
CASHEW, OIL-ROASTED, NO SALT	1 cup (129 g)	39 g	4 g	6 g
PISTACHIO	1 cup (128 g)	35.8 g	13.3 g	9.8 g
MIXED NUTS	1 cup (137 g)	34.7 g	12.3 g	

Source: www.Calorie-Count.com, May 2005.

When reading these Nutrition Facts, note that the sum of dietary fibers and sugars does not always add up to the total carbohydrates. Starches, sugar alcohols, and other forms of carbohydrates are not included in the listing. Keep in mind that the coconut meat, walnuts, and pecans at comparable amounts have the least amount of carbohydrates when compared to other nuts people often eat: peanuts, cashew, pistachio and mixed nuts.

In a Nutshell

While cholesterol plays an important role in a heart-healthy diet—which is why regular blood tests and doctor's consultations are important—there are many other reasons that someone could be prone to a stroke or heart attack, including Dr. Dayrit's Four I's of human disease: Inheritance (genetic), Infection, Immunology, and Inflammation (85). The first, genetics, is something that cannot be controlled by diet. But the last three are all responsive to the right diet, which includes the coconut because

- coconut oil is *cholesterol free;*
- coconut oil is *low in carbohydrates;*
- coconut oil is *free of hidden carbohydrates;*
- coconut oil is *trans fat free;* and
- with its *low linoleic acid* content, coconut oil does not make arachidonic acid-derived prostaglandins that may make blood prone to clotting.

Looking closely into the studies of the coconut, we find powerful evidence that . . .

- populations from coconut-eating countries are *heart-healthier* than those from non-coconut-eating countries;
- the incidence rates of coronaries soared to their current heights after coconut oil was virtually removed from American diets; and
- when eating coconut oil, the good HDL cholesterol goes up while the bad LDL remains the same.

With coconut oil, you can become leaner yet heart-healthier, and you won't be zapped by those free radicals. And there are still more benefits to the coconut!

Chapter Four

Coconut Oil May Help Protect Against Cancer

In This Chapter

- ☑ Cancer Research . . . Help Me Understand What's Going On
- ☑ Studies Show That Diets High in Polyunsaturated Fat May Promote Cancer
- ☑ Cancer-causing Versus Cancer-preventing Fats: The Omegas and Linolenic Versus Linoleic
- ☑ So-fine Omega-9 And Medium-chain Magic, Continued!
- ☑ More on How Coconut Oil May Help Prevent Cancer
- ☑ In a Nutshell

Cancer strikes fear into the hearts of most of us, and unfortunately, it seems that we know more and more people who have it or who know someone who has it. Whether you hear this about a friend, a member of your family, or from your own doctor about yourself, the news is scary. You know that cancer significantly disrupts one's life, is almost always disabling, and often it is a killer. As our fear grows, we try to learn more: what to eat, what some of the new treatments are, how to prevent it. This is laudable; as a physician, I always tell my patients to read up on their particular health concerns or on the research available about their specific condition. However, all the information out there—online, in the media, from friends, or as urban legends—can be confusing. About the importance of diet for promoting or preventing cancer, for example, I myself have read "do not eat eggs" then "eat more eggs"; "eat more fruits, vegetables, and low-fiber foods"; "take primrose and flaxseed"; eat "omega-3 oils" . . . or was it omega-6?

The reasons for following these various supposedly cancer-preventing dietary tips are often vague at best. Therefore, we tend to stay with a regimen for a while, until we slip back into our regular eating patterns or another dietary tip comes our way that sounds more plausible.

In this chapter, I ask you to join me as detectives on a case, following the major clues (meaning, the most reliable and constant conclusions) that have come from all the research thus far, in order to help you come to your own conclusions about what you can do in your diet to help prevent cancer as best you can.

Cancer Research... Help Me Understand What's Going On

What causes cancer? Why does it seem to strike more people nowadays? Is there a way to avoid it? Is it something in the diet? Can coconut oil reduce the risk of cancer?" Let me try to answer your questions, based on the large amount of research available.

Numerous studies have tried to understand why, out of a vast sea of similar cells, a group of cells appears that is distinctly not like the others. These cells look differently, act differently, and then—what's worse—

Rx: Coconuts! (The Perfect Health Nut)

they begin to grow in large numbers. This growth goes on independently of your body's controls for the normal regeneration of cells. In no time at all these renegade, now cancerous cells take over the precious control you have over your body's natural resources.

Fortunately, these studies have now identified some of the factors that influence why and how some cancer cells develop. The bad news is that you have no control over some of these factors. One good example is the genes your parents gave you. The good news is that there are other factors you *can* control: lifestyle, physical exercise, environment, even viral infections and, especially, *your diet*. It is this aspect of cancer prevention—what you eat—that we will concentrate on here, as it is well within your control and easy to put into practice.

"If countless studies have been done," you may ask, "why are there still so many who develop cancer?" The answer is simply that cancer research, like all scientific research, is like a giant puzzle board made up of the results of studies that ask one specific question at a time. The result of each study addresses not the big questions about cancer—what causes it and how we prevent it—only the specific question that was asked by the study. This new study then becomes another piece to be added to the puzzle board. Every now and then, other researchers examine a collection of these various puzzle pieces to try to understand the bigger picture of the root causes of cancer.

Putting these puzzle pieces together is very much like an investigation of a crime, in which one tries to solve a mystery by scrutinizing the clues. Let me guide you through the fine detective work that researchers have been doing through the years with regards to the relationship of diet and cancer. The clues are as enthralling as those on the popular television crime investigation series, *CSI*.

We'll probe into the studies that have been done in the complex world of *fat in your diet* and cancer research. As we act like those *CSI* investigators, we'll treat every new thing we learn from these studies as clues. Then, you can think about how they apply to you in particular. Once you understand *which*, and *why,* certain oils may promote, prevent, or even stop, the progress of cancer, you can solve the mystery of your own diet and really take charge of your food selections. Best of all, you'll learn about fats and oils in foods and how they can potentially help you protect yourself against cancer. You'll understand the different advantages and disadvantages of different oils, and will therefore find it easier to select what's right for you and to stick to any diet you choose. If you are ready to find out for yourself what oils to eat and avoid, and why, we can start with the first clue.

Clue #1: High-fat Diets Seem to Cause Cancer... But Certain Fats Seem to Prevent It. In studies on animals, more types of cancers developed on a high-fat than a low-fat diet. More than sixty years ago, researchers saw that certain chemicals stimulate normal cells to become cancer cells in laboratory animals (87, 88). They named these chemicals *carcinogens*. Next, they observed that the growth of the tumors (stimulated by carcinogens) was affected by the food given to the laboratory animals. They looked at several kinds of food and found that the growth of these tumors was affected by the oil in the animals' diet. Next, they looked at the amount of the oil and asked, "Would the amount affect these growths?" The results were as follows.

The researchers initially were surprised to see that the animals fed a diet high in fat developed more cancer growths. Conversely those who ate less fat had fewer of these cancer growths. One after another, the studies confirmed these findings in lab animals given carcinogens to grow cancers of the skin, liver, breast, colon, or prostate. These results were seen not just in one or two studies, but in most of them (89-94).

With regards to skin cancer, the investigators found the relationship to be true for both human and animal subjects. Skin cancers developed in the skin of animals exposed to ultraviolet (UV) light in the laboratory and in people exposed to outdoor sun. Given a high-fat diet, they develop more skin cancers (95, 96).

Your own dermatologist has given you the no-nos about sunbathing. This knowledge is so well established that even other specialists and the media have joined forces, telling you to avoid the sun or risk skin cancer or photoaging—the changes in the skin from sun damage. This means not just wrinkles but also the development of skin cancer. After my lectures, people ask me more questions about how to get rid of wrinkles and photoaging than about skin cancer. But the greater worry should be this:

the appearance of wrinkles and photoaging is a sign that cells have been damaged by the sun, and this can lead to skin cancer.

Given a high-fat diet, animals exposed to UV light developed more skin cancers. The relationship of cancer and the amount of fat in the diet of people was also examined, in a twenty-four-month population study. The results of this evidence-based study were: the volunteers who ate a diet high in fat grew more sun-induced skin tumors, whereas those who ate a diet low in fat grew fewer (97).

This is how the study was done: Patients were chosen who already had skin cancers of the types that dermatologists recognize to be due to past exposures to the sun. These patients were chosen because once one develops any sun-induced skin cancer, one is likely to develop more. To avoid bias, the researchers randomized the way these patients were accepted into the study, and likewise randomly assigned them to either of two groups: one group would eat a high-fat diet, the other, a low-fat diet.

At the time of the report, seventy-six patients had completed two years of the study. Thirty-eight of them had been told to continue their usual diet, which on the average had a high of 39% to 40% fat, never going below 36% fat. The other thirty-eight also started out with a similar diet that was high in fat. After they attended eight weekly diet classes they were able to switch to a low-fat diet. The results of these classes were truly impressive. One of these days I plan to find out how they instilled in these volunteers the motivation to get the following fantastic results: by the fourth month, the volunteers in this group were able to decrease their fat intake from 39% to just 21%! Even more impressive, they maintained this level throughout the rest of the twenty-four-month period.

On this lowered fat diet, their weight decreased during the first eight months, after which their weight remained about the same. At four-month intervals, dermatologists (who were "blind" regarding who belonged to which group) examined the sun-exposed areas of the face, neck, head, arms, and hands, looking for the earliest sign of skin cancer: thin somewhat reddish, rough, scaly spots called actinic keratoss.

From Months 4 through 24, the total number of new, early skin cancers—the actinic keratoses—found in the patients from the high-fat group was about ten; in those from the low-fat group, the number was about three. When the results were analyzed, the figures were found to be statistically significant. That means the results were reliable and not just due to chance.

Therefore, in both laboratory animals and in people on a *low-fat diet,* significantly *fewer* sun-induced types of early skin cancers developed, whereas on the *high-fat diet, more* skin cancers appeared.

Studies Show That Diets High in Polyunsaturated Fat May Promote Cancer

Continuing with your role as *CSI* agent—and recalling discussions from other chapters that showed you just how different one oil can be from another—you may start to wonder, "Which of the fats in those high-fat diets made those cancers develop? Is it all of them or is one fat more guilty than the others?" The scientific "detective-researchers" had the same questions, and so more studies followed, to narrow down the number of suspect fats. These studies lead you to this clue:

Clue #2: The Animal Studies Show That Polyunsaturated Fats Cause Cancer, and Coconut Oil Does Not. In 1987, Sylianco-Lim reviewed more than forty years of cancer research studies comparing the effect of coconut oil versus unsaturated oils. The verdict: coconut oil is cancer-innocent. The review findings were consistent. In the studies of animals stimulated by carcinogens, the animals fed coconut oil for six months produced less than half to no cancers at all versus those fed polyunsaturated oils for the same period of time. The latter group of animals developed many more cancers and the effect was dose related. As the amount of polyunsaturated fat in the diet was increased, more cancers grew; fewer cancers grew when the amount of coconut oil was increased (98).

This dose-related link between polyunsaturated fats and cancer growths has been seen in animals that have developed liver cancers from corn oil rich in polyunsaturated fats. Besides liver cancers, the animals fed corn oil developed gross anatomic changes of the liver, called nodular cirrhosis. In the other animals, regular and even hydrogenated coconut oil in the diet inhibited cancer growths. In fact, the cancer growths and nodular cirrhosis were kept down to as low as zero (99-101).

Another study, using several kinds of dye carcinogens to develop cancers of the colon and small intestines, corroborated this conclusion. Again, coconut oil had the same inhibiting effect on the cancers (102, 103). When the animals were fed not coconut oil but polyunsaturated fats—rich corn oil and safflower oil—many more tumors grew.

Yet other studies dealt with a cancer that strikes fear among us women: breast cancer. Here are some fascinating results about coconut oil, polyunsaturated fats, and breast cancer. Young female rats, given carcinogens to stimulate breast cancers and fed a diet of polyunsaturated fats from corn oil, developed more breast cancers than those fed coconut oil. As in the colon cancer studies, this was dose related. When changed to a combined diet of much less corn oil and more coconut oil, the breast cancers were fewer (104-107).

In still another study, a different set of carcinogens was developed to induce cancer of the pancreas in lab animals. The results were the same: animals fed coconut oil developed fewer cancers of the pancreas, and more when they were fed polyunsaturated fats-rich corn oil. As in the other studies, this effect was dose related (108).

Since the 1940s, the results of these comparative studies done on animals had become fairly consistent. By the 1980s, the tumor-growing potential of the polyunsaturated fats and of high-fat diets—but not of coconut oil—was widely accepted among medical researchers (109).

We'll go on to the studies in humans, but to understand them better, let's do another chemistry 101 review on the unsaturated fats, specifically with regards to the omegas, linolenic and linoleic acids.

Cancer-causing Versus Cancer-preventing Fats: The Omegas and Linolenic Versus Linoleic

Again, to better understand the potential impact of fats on cancer we'll need a little chemistry. If you prefer, skip ahead to the *In a Nutshell* section at the end of this chapter. If you'd like to get all the details of what could or could not help you prevent cancer, read on.

CANCER-PROBABILITY AND FATS: 101

Omegas, Linolenic, Linoleic. Like those intrepid crime investigators, let's get even more specific and ask, "Which fats in my diet are the guilty cancer-forming polyunsaturated fats? Are all polyunsaturated fats the same in terms of cancer-causing probability?"

The Omegas. Here the plot begins to thicken, because there are many kinds of polyunsaturated fats in your diet. You can understand them better by their family groupings, called the omegas. Just like in any human family, most of the family members are good, but some are not so good. And like the "bad egg" in most families, often there is a reason for why they become bad.

Before going into the omegas, stop to remember that all fats and oils are made up in part of fatty acids, which are composed of chains of carbon atoms. Remember, too, that polyunsaturated fats all have at least one double bond between their carbon atoms (saturated fats have only single bonds). Chemists love Greek words—remember *beta*-oxidation? Now comes omega, the *last* letter in the Greek alphabet. Chemists use omega to refer to the *last* carbon in the fatty acid chain. In the English language, we read from left to right. Chemists count from right to left. So, the last carbon is H_3C in the following example:

$$H_3C\text{-}C\text{=}C\text{-}C\text{-}C\text{-}C\text{-}C\text{-}C\text{-}C\text{-}C\text{-}C\text{-}C\text{-}C\text{-}C\text{-}C\text{-}COOH$$

Polyunsaturated fats fall into different omega groups, such as omega-3, omega-6, and omega-9. To see which omega group a polyunsaturated oil would fall under, we take the last, or omega, carbon (on the left, remember) and count the number of carbons (going right) to its first double bond (which looks like this: **C=C**). So, in the family of omega-3 polyunsaturated oils, the position of the first double bond is found at the third carbon from the omega (last) carbon. Similarly, counting from the omega carbon, the sixth carbon is the site of the first double bond in the omega-6. And again, for the omega-9 oils, the first double bond is seen at the ninth carbon.

Rx: Coconuts! (The Perfect Health Nut)

Here are the chemical signatures of three common 18-carbon chain fatty acids:

H₃C-C-*C=C*-C-C=C-C-C=C-C-C-C-C-C-C-C-COOH   Linolenic acid   An omega-**3**
H₃C-C-C-C-C-*C=C*-C-C=C-C-C-C-C-C-C-C-COOH   Linoleic acid   An omega-**6**
H₃C-C-C-C-C-C-C-C-*C=C*-C-C-C-C-C-C-C-COOH   Oleic acid   An omega-**9**

The omega families of polyunsaturated fats in your diet are: *omega-3, omega-6,* and *omega-9*.

The names of some important *omega-3* polyunsaturated fatty acids are linole*nic* acid, EPA, and DHA. These are names that are new to you, but I mention them now because they are considered the *good* oils, and we will talk more about them later.

The principal *omega-6* polyunsaturated fatty acid is linoleic acid. This acid you already know as the somewhat problematic one from Chapter 3 (too much *linoleic acid* can promote arachidonic acid-derived prostaglandins that may make blood prone to clot). In this chapter, you will see how linoleic acid also can promote cancer growth.

First, let me help you get more familiar with the omega families of the fatty acids that are in your diet.

➢ The principal **omega-3** polyunsaturated oils are EPA and DHA, which are abundant in oils from fish, and in the flesh of fatty fish. Another omega-3 is alpha linole*nic* acid, which is abundant in flaxseed (about 60%) and in green, leafy vegetables.
➢ The principal **omega-6** polyunsaturated oil is linoleic acid from vegetables/seeds: regular safflower (78%), sunflower (68%), corn (57%), cotton (53%), and soybean (53%).
➢ The principal **omega-9** oils are monounsaturated and can be found in oleic acid—rich in walnut, macadamia, olive, peanut, hybrid safflower, and sunflower; also chicken, duck, goose, turkey fat, and lard.

Linolenic Acid Versus Linoleic Acid. To help you remember linole*nic* and linole*ic* acid—acids with two very *similar* names but with *opposite* effects—try this:

-Eic ➡ ICK ☹ ➡ Sick Omega **SIX**

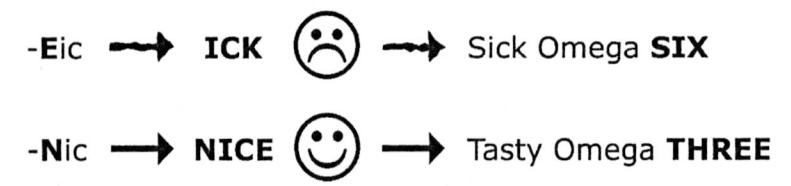

-Nic ➡ NICE ☺ ➡ Tasty Omega **THREE**

To further help you distinguish them further on, I've written their names to highlight their differences: linole*Nic* acid and linol*Eic* acid. How important are these omega names in trying to solve the mystery of why cancers develop? Very. And just like real estate, it's got to do with location, location, location.

Believe it or not, location is as important to the effect of these fatty acids on your health as it is to a store's financial health when it decides where to open a new outlet. The location of that polyunsaturated fat's first double bond makes each omega family different from the others when it comes to how they can increase or reduce the risk of cancer.

- Linol*Eic* acid, an omega-6, has been shown to *enhance* the growth of cancers.
- Alpha linole*Nic* acid, EPA, and DHA are omega-3 and have been shown to *decrease or prevent* the growth of cancers.
- Oleic acid, an omega-9, has also been shown *to decrease or prevent* the growth of cancers.

Let's look at these clues from studies using the animal models in the laboratory. The results again are consistent: in a substantial number of studies, animals with omega-3 polyunsaturated fats in their diet had fewer or no cancers whereas those with omega-6 polyunsaturated fats had more cancers (110-120).

Although the animal studies were fairly conclusive, when people were studied, the results were not. The hypothesis that omega-3 *diminishes* while omega-6 *enhances* the growth of cancers has not been proven consistently true across all human studies. The association was seen in some small clinical studies and even when whole groups of people were studied (121-131). However, the rest of the studies reported this association to be inconclusive (132-145). Three recent analytical papers found that only about one-third to one-half of the studies showed this association to be present (146-148).

Why is this so? Let's consider the explanations for the inconclusive findings in human beings, proposed by the authors of these analytical papers. For starters, in the laboratory it is quite easy to control animal subjects. One gives them their food, weighs them, and observes them for the appearance of tumors. No problem. However, population studies deal with people like you and me. I don't know about you, but certainly most human beings I deal with (myself included) are not the most cooperative test subjects. Despite the investigators' best efforts to account for all these differences, errors are bound to happen in studies like these because they often rely on diet recollections on questionnaires or in a diary. Can you remember what you ate for breakfast, lunch, or dinner on any given day? I can't even remember where I placed my reading glasses a few minutes ago!

This recall flaw can be an even greater problem if a test subject is asked to go further back in time, which is important. The findings from animal studies indicate that looking at multiple generations is critical. Diet exposure during pregnancy was found to affect not just the pregnant animal but also its offspring and even their offsprings' subsequent offspring (149-151).

Cultural eating differences among human test subjects was proposed as another reason for the inconsistent study findings. People who generally eat a lot of fish have been shown to have a lower rate of cancers of the breast, ovaries, and prostate (152-154). This is true for the Japanese (155-156), whose fondness for fish many of us have adopted, with our gravitation toward sushi and sashimi bars. Interestingly, in recent years the Japanese ate relatively less fish and more vegetable oils rich in omega-6 polyunsaturated fats. Subsequently, Japanese women have seen increased breast cancer rates (157-158).

Like the Japanese, Swedish people tend to eat more fish than most people. Similar findings of low incidence of cancer was seen in Norway (159), in a nationwide Swedish study (160), among Eskimos (161) and among people from Greenland (162). The higher baseline level of omega-3 among these fish-eating people is believed to explain the difference between the cancer rates among them and the rate among those who eat fish to a lesser degree.

Look in your medicine cabinet for still another possible reason for the inconsistent studies among human volunteers. A review of complementary and alternative medicine usage has shown that more and more people are taking many kinds of supplements (163). Included in the list are antioxidants such as vitamin C and, increasingly, newer plant antioxidants. These antioxidants are also added to prepared food products to enhance their perceived value to consumers, or even just to help make the product stable. These and equally common over-the-counter headache, pain, and anti-arthritis drugs have been shown to lessen the cancer-protective effect of omega-3 polyunsaturates (164, 165).

I have just shared with you the possible reasons why the animal studies were quite consistent but those done on people were not. The authors of these analytical papers confirmed these conclusions, after a review of numerous studies. They looked at scientific literature, not just from one country but also worldwide. They included large population groups and studied the food associations back in time as well as into the future. Besides animal studies and population studies with people, they included studies that use modern scientific laboratory tools on the fats and oils in our tissues. The authors included only those studies that had the greatest statistical validity and the least bias.

Having looked at these studies in a similar manner, I am convinced by these authors' analysis. You now have a choice: reject the reasons they propose, or accept them. If you are not convinced about the coconut as a food that can protect against cancer, go on to the next chapters to read about the coconut's many other uses as a high-fiber flour (higher than wheat, rye, or oatmeal), as a source of growth hormones, as an antibiotic, and as a moisturizing and antioxidant oil for your skin.

Should you choose to accept their analysis, continue with your investigation and move on to the next clues.

Clue #3: Omega-6 Linol*Eic* Acid Can Promote Cancer. Omega-3 Linole*Nic* Acid Can Protect Against Cancer.

In a number of multinational studies, biopsies of fatty tissue taken from patients were measured for the ratio of omega-3 to omega-6 polyunsaturated fats. The greater the amount of omega-3 found in the fatty tissue, the fewer the breast cancer growths

(166-168). If you accept results such as this, as well as the conclusive animal studies and the qualified conclusions of the human studies, you ought to eat *more* alpha linole*Nic* acid (-Nic = nice = tasty three) from flaxseed and more EPA and DHA from fish (the oilier, the better). Do you like to eat herring, sardines, mackerel, or salmon? If you do, be happy. Their fish oil content is among the highest. The leanest are cod, sole, grouper, and haddock. In between those two in oil content are freshwater bass, carp, catfish, and tuna (7). On the flip side, you ought to be eating *less* linol*Eic* acid (-Eic = ick = sick six) from soybean, corn, cottonseed, and sunflower.

When we say we ought to eat more or less of these oils . . . just how much are we talking about? More recent studies on the *amount* of fat now appear to show that, in the production of cancer, it is not *how much* per se, but the *ratio* of omega-3 to omega-6 polyunsaturated fats that counts. (On the problem of getting obese, however, amount *does* count; we'll talk more about that at the end of this chapter).

Clue #4: Pay Attention to the Omega-3 to Omega-6 Ratio. To reduce the risk of cancer, the ratio of your intake of omega-3 to omega-6 should be **1:1** or **1:2**.

The recommendation of this ratio is based on diets that are known to be associated with a low cancer incidence in specific population groups: ancient Paleolithic humans (169), the classic Mediterranean diet (170), and the Lyon diet (171). The populations that eat these kinds of diets are known to have very low rates of cancer. Their ratio of 1:1 to 1:2 comes from the eating of omega-3 from oily fish and omega-6 oils from vegetable and seed nuts.

While this may appear to be the kind of diet that you and your family eat, be careful. As you may recall from Chapters 2 and 3, you probably eat much more of the linol*Eic* acid-containing polyunsaturated fats, because they are often *hidden in your diet.* Many of these polyunsaturated fats are partially hydrogenated and rich in trans fats. Because they have been rendered stable by hydrogenation, they end up everywhere: in food cooked at home and at restaurants, in processed and prepared foods. As we become more industrialized, we exercise less and eat fewer complex carbohydrates, fruits, and unprocessed vegetables. We also tend to eat a lot more refined or processed foods and snacks, as well as low-fat foods that have hidden polyunsaturated fats and trans fats.

This means we tend to consume *less* omega-3 (-Nic = nice = tasty three) but *more* (albeit, perhaps hidden) omega-6 (-Eic = ick = sick six) and trans fats. Remembering that the ideal ratio of omega-3 to omega-6 intake is 1:1 or 1:2, it may shock you to learn that in most developed countries, the ratio of omega-3 to omega-6 ranges from 1:10 to 1:20 (170)! And often, this ratio reflects the absence of coconut oil in the diet. To repeat: coconut oil and omega-3 oils reduce cancer growths, whereas omega-6 oils stimulate cancers.

So far so good . . . the key phrases up to this point are:

Linol*Eic* acid	=	-Eic	=	ick	=	Sick omega-**6**
Linole*Nic* acid	=	-Nic	=	nice	=	Tasty omega-**3**
Omega-3 to omega-6 Ratio				=		1:1 or 1:2

But is that all? Hardly. Besides the good linole*Nic* acids from tasty omega-3s, there are other oils that you'll be happy to know can help prevent the risk of cancer and will make delicious additions to your diet.

So-fine Omega-9 . . . and Medium-chain Magic, Continued!

We are now going to add the following to your key phrases:

| Oleic acid | = | So-fine omega-**NINE** |
| Medium-chain saturated fat | = | Mmm**Magic** fatty acids |

Clue #5: So-fine Omega-9 Oleic Acid. At this point, you're probably wondering about olive oil. Rightly so, as olive oil is widely talked about as a "good" oil. In fact, oleic acid, an omega-9 found in olive oil, has been shown in animal studies to *reduce* or to *not produce* cancer (98).

Go back to our chart in Clue #2. It shows that oleic acid is an omega-9 and is a monounsaturated fatty acid. Studies have shown that, like the omega-3 polyunsaturated fats, the omega-9 monounsaturated fats reduce and may even protect against cancers. One such study, on rats, showed that a corn oil diet stimulated more cancers than a control diet, whereas olive oil led to fewer and smaller tumors than the control diet (172).

In recent years, cancer cells have been made to grow in the laboratory. A very recent study on laboratory-grown breast cancer cells showed that oleic acid reduced levels of the gene Her-2/neu by up to 46%. The significance of this finding is that in more than one-fifth of people with breast cancer, there are many more of these genes in their cancer cells. Its presence appears to signal a particularly bad kind of breast cancer. This study also showed that oleic acid increased the expression of a protein that works to suppress the tumors and enhanced the effectiveness of the drug Herceptin, which targets the Her-2/neu gene (173).

That monounsaturated fat-rich olive oil helps protect against the development of cancer has been confirmed in a series of human studies (174—180). One study of 61,471 women in Sweden found that monounsaturated fats protected against breast cancer, polyunsaturated fats increased the risk, and saturated fats were neutral (175).

Also, where olive oil is a staple, in places such as Italy, Spain and Greece, the incidences of breast cancer are lower than in North America and Northern Europe, both of which consume less olive oil (176-179). A study on the relationship between the fatty acid content of fat tissue and breast cancer likewise shows a strong inverse association: the higher the oleic acid in the fat stores, the lower the number of cancers (180).

As I've mentioned, oleic acid is abundant in olive oil (71%), but it also is found in high amounts in oils from the almond (61%), avocado (51-68%), hazelnut (69-81%), and rapeseed (33%). Walnuts have 23% oleic acid, but they also have a high (54%) linol*Eic* acid content (-Eic = ick = sick six).

People often associate olive oil with healthy diets, and that usually means no saturated animal fats, which are known to increase the bad cholesterol (LDL). It may come as a surprise to you, therefore, to learn that the oleic acid of olive oil is also present in goose (56%), eggs (50%), and the fat of chicken (42%), duck (46%), and turkey (38%). This is nice to know, isn't it? Some of the animal fat you eat has cancer-preventing oleic acid, too.

As a good detective, you've probably noticed that we still haven't gone into detail regarding the cancer impact of a type of oil that featured heavily in our previous discussions, medium-chain saturated fats. Having saved the best for last, this is now our final clue.

Clue #6: Medium-chain Magic, Continued! As you may have guessed by now—given all the other benefits of medium-chain fatty acids that we've discussed in other chapters—diets containing medium-chain saturated fats (which, again, are plentiful in coconut oil) have been shown to prevent cancer.

In the animal studies that compared olive and coconut oils, both reduced the number of tumors, but the tumors were lowest, often down to *zero*, in the animals fed coconut oil (98).

In addition, while olive oil has been shown to prevent the promotion and spread of cells that are already transformed into cancer cells (181-182), coconut oil has been shown to act even *earlier*. In little-known, though published, studies by Lim-Sylianco, coconut oil was found to act at the initiating phase of cancer growths. She found coconut oil to be antigenotoxic, which indicates that it actually helps prevent the *formation* of cancer cells (183-184).

And this is just the beginning of what coconut oil can do in the area of cancer prevention through a healthier diet.

More on How Coconut Oil May Prevent Cancer

Just why and how does coconut oil prevent the formation of cancer cells? Earlier I said that you needed to know the answers to *why* and *how* in order to want to stick to a cancer-freeing diet. Let's continue to follow the work of these detective researchers that explains the why and the how of coconut oil (146-148).

Clue #7: Coconut Oil Creates Healthier Cell Membranes. Quite simply, cell membranes are like the handbags (or tote bags or backpacks) of your cells. You need strong cell membranes to keep the cells healthy. Like

Rx: Coconuts! (The Perfect Health Nut)

a woman with many handbags, the body has several handbags, in all sizes, for every imaginable need. One large handbag is the *plasma cell membrane,* which bags each individual cell to separate it from all the other cells. Then, the contents of a cell themselves are also neatly surrounded by smaller handbags (or wallets or cell phone holders) to separate them from one another (to keep them organized, as it were). The contents—and the handbags that surround, organize, and protect them—are as important as the keys, makeup, sunglasses, and wallet that you carry around in your real carryall, though they are definitely more vital to your existence.

These baglike membranes are made up of mainly phospholipids and cholesterols. Phospholipids have two main component molecules: a phosphate and a fatty acid, which can be a

Polyunsaturated fat (PUFA)	H3C-C-C-C-C-C=C-C-C=C-C-C-C-C-C-C-COOH	e.g., linol*Eic* acid
Monounsaturated fat (MUFA)	H3C-C-C-C-C-C-C-C-C=C-C-C-C-C-C-C-COOH	e.g., *o* leic acid
Saturated fat	H3C-C-C-C-C-C-C-C-C-C-COOH	e.g., lauric acid

The phospholipids mostly come from the oils you eat. Some of the cholesterol component comes from the food you eat while the rest is made by your liver in response to your body's needs (185).

You may recall from Chapter 1 the difference in the texture of coconut oil and the other saturates when compared to the polyunsaturated fats: the saturated fatty acids (and cholesterol) are firm fats; the polyunsaturated fatty acids are thin liquids.

You're probably familiar with how a dish of Jell-O is made, so let's use it to demonstrate what happens at your cells. At a correct ratio of water and gelatin, a dish of Jell-O comes out just right: neither too floppy nor too firm. Likewise, at a correct ratio of the liquid polyunsaturated fats + the more dense saturated fatty acids + cholesterol, the cells' membranes have the proper balance of rigidity and flexibility, which allows them to function well.

The current diet recommendation we have been given by nutrition and diet experts is low fat . . . but remember that this often equates to a diet that is high in polyunsaturated fats. To make a diet low fat, the tendency is to replace with polyunsaturated fats the saturated, mostly animal and dairy fats in the usual diet. In the skin cancer study, for instance, the polyunsaturated fats increased from 60% to 80% in those who were given a low-fat diet.

Do you remember the example of the low-fat diet in Chapter 2, and how the decrease of saturated fatty acids (from coconut oil, for example) and the tremendous increase in unsaturated fatty acids becomes reflected in the fat-to-oil ratio of your cell membranes? It is the same case here: since we tend to eat a lot more of the more liquid polyunsaturated fats (especially on low-fat diets), the balance of fats needed by your cells' membranes is skewed toward a more fluid, less firm cell membrane.

This imbalance—a less firm cell membrane due to too many liquid unsaturated fats—is also seen at the cell membranes' antennae-like receptors. With these receptors, your cells are like your mobile phone, allowing circulating hormones to signal your cells what to do. The change of membrane fluidity or rigidity may alter your cell's receptors, resulting in altered responses to signals to and from other cells as well as within the cell itself. The garbling of these signals, like the crossed signals over a mobile network, may eventually lead to the development of disharmony, or poor cell interactions. These alterations begin to transform previously healthy cells into the enemy: cancer cells.

Let's look again at those laboratory animals in the studies. The animals that develop colon cancers while on a high-polyunsaturated fats diet have been shown to have high amounts of the cancer-promoting omega-6 (-Eic = ick = sick six) polyunsaturated fats. These are found both in the membranes of the cells that line the colon and in the colon cancers themselves (103).

Your cell membranes are also important for the cells that handle the immune responses of your body, the *lymphocytes*. Lymphocytes are very common cells. If you read a blood-test report of what doctors call your complete blood count, you can see that lymphocytes are listed there. These are the cells that are vital for the immune functions of your body. Among these lymphocytes is a special one you may have heard of, called the T cell.

These T-cell lymphocytes are your guardian cells against anything foreign, including bacteria and viruses, chemicals, or even sun-induced cell damage. They naively approach these foreigners in our body and check them out. Once they know them and find them objectionable, their memory is phenomenal. Every time the T cells meet up with the intruders again, they mount a reaction against them. This is how a reaction happens—say, an allergy to a drug, a fever, or an infection.

T-cell responses have been shown to be markedly *reduced* in those on high-polyunsaturated fats diets. A detrimental effect on our body's surveillance system, this change is also believed to be due in part to changes in the fluidity of the membranes of the T cells (180).

In summary: by virtue of their nonsaturation, oily polyunsaturated fats at the membranes adversely affect membrane fluidity and the signaling of cells. Thus, normal cells may become cancer cells. By a similar deleterious effect on the fluidity and signaling of the guardian T cells, cancer cells are allowed to evade the immune system. In this way the cancer cells are able to spread to distant parts of the body. Increasing your intake of coconut oil—because it is a thicker saturated fat—helps to maintain the crucial balance of your cell membrane's fluidity and rigidity.

Clue #8: Coconut Oil Does Not Create Free Radicals That Harm Cell Membranes. Remember our lesson on double bonds in Chapter 2? Coconuts are saturated, have no double bonds, and are therefore more stable and are not readily oxidized. The oxidation of the unsaturated fatty acid double bonds of polyunsaturated fats produces *reactive oxygen chemical species* and other free radicals that are damaging to cells, the results of which are a prelude to cancer. The monounsaturated fats, and especially saturated coconut oil, do not produce this problem.

Normally, your cells are able to repair this damage, using natural antioxidants and other inherent repair mechanisms built into cells. However, in the presence of too many free radicals from too much polyunsaturated fat, the cells lose the fight. In time, they are no longer able to return to their normal, healthy state. They become cancer cells.

Clue #9: Coconut Oil Does Not Adversely Affect the Balance of the Tissue Hormones at Your Cells, whereas linol*Eic*, DHA, EPA, and even otherwise nice linole*Nic* acid all do.

Remember that linol*Eic* acid forms arachidonic acid, from which are made more of the *pro*-inflammatory tissue hormones. EPA, DHA, and linole*Nic* acid form more of the *anti*-inflammatory tissue hormones.

Coconut oil is not involved in the making of these tissue hormones.

What are tissue hormones? Hormones, in general, are chemicals produced by specialized cells, to stimulate other cells to become active for specific jobs. Have you heard of insulin or thyroid hormones? Those two are hormones produced from the fatty acids of cells in specific organs of the body. The hormones are carried by blood vessels to their target cells.

A fundamental law of nature is balance. At the level of your cells, the tissue hormones are formed to help maintain this balance (186). The omega-3 and omega-6 fats produce several tissue hormones that have competing actions or that are antagonistic to each other. Their opposing actions keep your cells in balance.

The omega-6 linol*Eic* acid is converted to arachidonic acid and then into growth hormones that are largely *pro*-inflammatory. That means they can incite changes so tissues become red and swollen.

Providing balance, the omega-3 linole*Nic* acid-derived hormones compete with these effects and are largely *anti*-inflammatory.

When the ratio of the omega-3 to omega-6 is highly in favor of the omega-6, *pro*-inflammatory conditions arise. These can lead to inflamed tissues of specific sites, which can then manifest as some of the more common modern diseases of the lungs (asthma), of the fibers called collagen (connective tissue diseases), of the joints (arthritis), and of various parts of the skin (allergies, psoriasis) (187).

Now you can make a more knowledgeable decision about whether you need the following popular nutritional supplement oils:

- Primrose oil is said to be good because it has about 10% of another linole*Nic* acid, an omega-6, which is believed to be *anti*-inflammatory. Its problem is that the oil has more of the *pro*-inflammatory linol*Eic* acid, around 72%.
- Flaxseed oil, another nutritional supplement, has 60% of the *anti*-inflammatory alpha linole*Nic* acid, and its linol*Eic* acid content is only about 14%.

Rx: Coconuts! (The Perfect Health Nut)

Let's summarize again: cancers develop on a polyunsaturated fats-rich diet, especially one heavy with omega-6 oils. Having more polyunsaturated fats at the membranes of your cells makes them malfunction as a signal receiver, affects your T-cell lymphocytes, decreases immunity, and allows cancer cells to develop and grow—and they get oxidized, zapping at the less stable unsaturated double bonds of the polyunsaturated fats (188).

More linol*Eic* acid, the most common of those polyunsaturated fats, produces *pro*-inflammatory hormones that, together with oxidation reactions, may also allow cancer cells to develop.

In a Nutshell

After all these clues, what have we learned about cancer and the oils in our diet?

- **More** cancers develop in diets that are

    o high in total fat (Clue #1); and
    o high in omega-6 polyunsaturated fats (Clues #3 and 4).

- **Fewer** cancers develop in diets that are

    o high in omega-3 polyunsaturated fats (Clues #3 and 4);
    o a ratio of omega-3 to omega-6 polyunsaturated fats at 1:1 or 1:2 (Clue #4); rich in omega-9 oleic acid (Clue #5); and
    o high in the saturated *medium-chain* coconut oil. (Clues #6 through 9).

- Coconut oil helps protect against cancer (Clues #2, 6-9).
- Key phrases to remember are:

89

Linol*Eic* acid	>>	-Eic	>>	ick	>>	Sick omega-**SIX**
Linole*Nic* acid	>>	-Nic	>>	nice	>>	Tasty omega-**THREE**
Omega-3 to omega-6 ratio	>>	**Balance**			>>	1:1 or 1:2
O leic acid					>>	So-fine omega-**NINE**
Medium-chain saturated fat	>>				>>	Mmm**Magic** fatty acids

Do I have suggestions on how you can improve your diet to lessen your risk of getting cancer? Absolutely! In Chapter 11 (Rx 3), you'll find the Rx: Coconuts Lifestyle. In addition to showing you how to use the coconut for your skin and general health, its dietary recommendations are designed to help you stay slim, keep your heart healthier, and to possibly prevent and protect you from cancer.

Chapter Five

Coconut Monoglycerides: Impressive Results in Infectious Diseases

In This Chapter

- ☑ Brief History of Coconut Derivatives Used for Infectious Diseases
- ☑ Laboratory Studies of Coconut Lauric Acid Monolaurin by Dr. Jon Kabara
- ☑ How the Coconut's Triglycerides Become Monoglycerides
- ☑ Clinical Studies Using Monolaurin and Other Coconut Monoglycerides
- ☑ How Do Coconut Monolaurin and Monoglycerides Work?
- ☑ In a Nutshell

I first encountered Dr. Jon Kabara in 1998, at a dermatological meeting in Beijing. His talk focused on the use of a new (to me) coconut-derived antiseptic, monolaurin, which he had discovered in the 1960s, to create preservative-free cosmetics. Most cosmetics are preserved with formaldehydes, which tend to cause irritations among those with sensitive skin. The more allergy-aware cosmetics use parabens in small doses, but people with very sensitive skin still show sensitivity to parabens. Dr. Kabara's talk introduced me to the possibility of a nonirritating, natural, and safe preservative for cosmetics. My interest was piqued. Since then, through the studies I will share with you in this chapter and through my own experiences in treating patients with this coconut derivative, I have been constantly impressed with the variety (and success) of monolaurin's applications. From treating skin conditions to preventing viral and bacterial infections, this chapter will help you reap the full benefits of this truly amazing coconut derivative.

Brief History of Coconut Products Used for Infectious Diseases

Since ancient Ayurvedic times, healers have used many products of the coconut to treat their patients (189). They used coconut products as an antiseptic, an astringent, a gargle for sore throats, a dentifrice for curing toothaches and strengthening gums, and even for earaches and stomach aches.

Shown to decrease any fever, coconut products were also used to treat and prevent diseases such as scabies, syphilis, gonorrhea, smallpox, dysentery, leprosy, tapeworm and other intestinal parasites, ringworm, parasitic skin infections, skin ulcers, and boils. These infectious diseases are as old as biblical tales, but they still present problems for many people today (190-191).

It is anthropologically interesting to note that primitive healers and early physicians knew about these therapeutic effects without the benefits of medicine's current scientific protocols. Their knowledge came from employing time-tested native practices while heeding a healer's inner voice. Acting on this instinct, these treatments were found to deliver real results and then were passed on as a form of traditional medicine (192).

Considering the scientific means we now have at our disposal, it is rather a shame—especially considering the groundbreaking discoveries being made every day about the benefits of previously dismissed natural remedies—that modern physicians writing about these benefits of coconut oil are still few and far between (193-196). Worldwide, there have been only a handful of coconut-advocating or even coconut-researching modern scientists. Going against popular, often misguided medical beliefs, their passion in examining the safety and curative actions of the coconut has been endless, and maybe even . . . infectious!

Laboratory Studies of Coconut Lauric Acid Monolaurin by Dr. Jon Kabara

Foremost among these investigators of coconut products is Dr. Jon J. Kabara, a chemist with a Ph.D. in pharmacology, whose studies on fats and oils are broad and far-reaching. His discovery of the antiseptic action of the coconut monoglycerides in the 1960s was ahead of its time. While he has received acclaim, recognition of his revolutionary work is just recently on the rise, and widespread recognition in the medical profession has yet to be seen.

In 1966, the young scientist Dr. Kabara started his studies on the breakdown products of the fats and oils we eat. He was looking at them as possible sources of growth factors. Instead, he found that lauric acid and monolaurin produce the opposite result: they *stopped* the growth of bacteria (197). These surprising effects were to become a major passion for this multitalented chemist.

Through the 1970s and 1980s, in one study after another, he found that monoglycerides have antiseptic properties against certain organisms. He also showed that the antiseptic potency of these monoglycerides varied, depending on the specific fatty acid attached to the glycerol. He found lauric acid to be the fatty acid with the best antiseptic ability. In still more studies, he found monolaurin, the monoglyceride of lauric acid, to be even more effective than lauric acid alone (198, 199).

This is the surprise: lauric acid is present in a lot of food oils; however, it is present in coconut oil and palm kernel oil (another tropical oil) in quantities *up to 50% of total volume*—far more lauric acid content than in all other oils.

Excited to find out more about monolaurin's antiseptic activities, Dr. Kabara studied its effects on several organisms and found it effective against fungi (200) and a range of bacteria (201-203). He then worked with scientists in the food and cosmetics industries, where monolaurin, an oily substance, is used as a base ingredient. Having discovered its antiseptic activity against bacteria and fungus, Dr. Kabara introduced to the cosmetics industry the dual action of monolaurin (which he patented under the name Lauricidin) as a basic product ingredient and as an antiseptic that could kill microorganisms in cosmetic products (204-206). For the food industry, he revealed the use of monolaurin's dual action as an ingredient and as an antiseptic for food products (207-208).

Now comes an even greater surprise. The impressive antiseptic activity of monolaurin against fungi, even bacteria, are good enough . . . but how about monolaurin's effectiveness against the more difficult-to-treat, and even more deadly, viruses? Working at the United States' Centers for Disease Control and Prevention, Drs. Hierholzer and Kabara first reported the antiviral activity of lauric acid and monolaurin on certain RNA and DNA viruses (209). Before we continue on this, let me first show you how coconut oil/meat becomes a monoglyceride.

How the Coconut's Triglycerides Become Monoglycerides

We'll start with the triglycerides.

You may recall from Chapter 2 that the oil in the steak you had for dinner breaks down into this basic unit. The prefix *tri*—indicates that there are three fatty acids in a triglyceride molecule. Each of these fatty acids is attached to glycerol, a simple and short chemical compound with just three carbon atoms.

I like to think of glycerol as the character Bosley in the TV series and movie *Charlie's Angels*. Imagine the three fatty acids as Charlie's three irrepressible (but not at all fatty!) Angels, each attached to the arms and a leg of Bosley to make up a *triglyceride*. Remember, too, that one of these fatty acids is the important lauric acid. Now imagine Bosley and the Angels doing a free fall from the stomach into the intestines. One fatty acid Angel slips off, leaving two fatty acids attached to Bosley, which now makes a *diglyceride*. Then another fatty acid Angel slips away from Bosley. One fatty acid is left attached to the glycerol. This is now called a *monoglyceride*.

Enzyme-derived Coconut Oil (EVCO). Two years ago, enzyme-derived virgin coconut oil (EVCO) and the coconut monoglyceride produced from it became available for me to study. These are produced by Dr. Teresita Espino at the University of the Philippines' Agricultural School, using food-safe butyl alcohol and a natural rice bran lipase. Compared to coconut oils produced from other methods, EVCO, has an even higher concentration of lauric acid: 76% versus the 50% found in other virgin coconut oils. In proportion to this increase, the other medium-chain fatty acids—caprylic and capric acids—decrease from 15% to 6%.

The coconut monoglycerides are derived from EVCO. These are slightly different from Dr. Kabara's Lauricidin. Lauricidin is made from the lauric acid (just that one fatty acid) of coconut oil, while Dr. Espino's monoglyceride comes from full coconut oil. Therefore, coconut monoglyceride has mostly lauric acid, plus any of the other potentially antiseptic fatty acids in the coconut oil—C_8, C_{10}, or C_{14}.

Presently, we have also started to compare monolaurin and the EVCO monoglycerides, in the hope of widening the treatment applications of the coconut's fatty acids' glycerides.

EVCO monoglycerides in these early clinical studies show promise of having Lauricidin's broad-spectrum antibiotic actions. The EVCO monoglycerides appeared to be similar or slightly superior in its antibiotic action, but these are preliminary data and more studies are needed to establish their effects and statistical validity.

Clinical Studies Using Monolaurin and Other Coconut Monoglycerides

Several other researchers began to take note of Dr. Kabara's findings, leading, since the late 1980s, to more studies of monoglycerides in the laboratory. These confirmed Dr. Kabara's early studies of monolaurin's wide-ranging anti-infective action (210) against bacteria (211-216), fungi (217), and viruses (218, 219).

The studies on viruses (209, 218-219) deserve special mention because, compared to antifungal and antibacterial antibiotics, there are only a few antiviral drugs. For those of you who remember the dreaded herpes scare in the 1980s, think of the few medicines that were available for that disease.

The list of viruses that are killed by monolaurin in the laboratory is impressively long (219). Included are common and not-so-common viruses: measles, herpes simplex, hepatitis C, viruses that produce mouth sores, and cytomegalovirus. This last virus is called *opportunistic* because in normal states it is fairly harmless but it makes victims of those whose immunity is lowered by lingering diseases.

Even more surprising was that monolaurin, in the laboratory, could kill the more deadly visna virus and HIV.

Clinical Studies. In the development of any drug, after *laboratory* tests and studies, lengthy *clinical* studies are always needed. It is a process that goes through several stages, called Phase 1 to 4 trials. These trials, done first on healthy people and then on those sick with a particular disease, using a few and then large numbers, are meant to assure the public of the safety of the product. Safety means the product will treat the disease without itself causing additional problems or significant side effects to the person being treated.

Monolaurin has not yet been subjected to a wide set of these clinical trials. After monolaurin was used as an antiseptic in cosmetics and food, many non-doctors started to become aware of it. Spreading by anecdotal experience and hearsay, people using alternative medicine products started to use "coconut monolaurin" for various problems that needed an antiseptic. Since then, its use has grown in leaps and bounds.

My last check of the Internet on April 10, 2005, came up with more than six hundred citations on Lauricidin and five thousand on monoloaurin. Just type "monolaurin" into Google and check them out for yourself. But because both monolaurin and Lauricidin are not yet a part of mainstream medicine, and to encourage you to rely on clinical and medical studies over hearsay or Internet popularity, it is time to recount some of the clinical trials using coconut lauric acid's monolaurin.

Clinical Study Number 1: HIV. This very first clinical trial for monolaurin studied nothing less than HIV, the virus responsible for AIDS. This clinical trial came about this way. Dr. Mary Enig, in a lecture on coconut oil at a 1999 symposium in Manila, mentioned this story from the AIDS organization Keep Hope Alive: In the Caribbean, several HIV-AIDS patients were no longer responding well to their medications. On the basis of the coconut's traditional use as an anti-infective, these patients were told to continue their medicines but to also eat half a coconut each day. Following this, their blood HIV counts were said to fall, in some cases, to undetectable levels (220).

In the audience that day was Conrado S. Dayrit, M.D., an emeritus professor in pharmacology, who initiated clinical studies on the pharmacological effects of coconut oil (54, 85). He had worked with Dr. George Blackburn and other researchers from Harvard and Columbia Medical Schools to prove that coconut oil deserves to be medically, pharmacologically appreciated (73). The fact that these researchers' findings were unnoticed was not their fault. As I mentioned earlier, controversial subjects stand a poor chance of being published in mainstream journals.

We physicians have a dismissive name for this kind of story: *anecdotal*. However, because of his own research experience with the coconut, Dr. Dayrit decided to follow up on Dr. Enig's story. In addition, Dr. Enig is a Ph.D. in nutritional sciences who is well known for her comprehensive analysis and review of trans fats. As a proponent for trans fat-free coconut oil in the diet, she was one of the earliest voices articulately alerting the public to the dangers of trans fat and at the same time promoting the virtues of coconut oil. Even if her story was anectodal, Dr. Enig was a reputable source.

Dr. Dayrit thought there might be merit to an open clinical study of HIV patients, using coconut oil and its purified monolaurin. At this point, neither monolaurin nor its parent coconut oil had previously been documented in clinical trials for systemic infections. In a way, this study was a shot in the dark for patients who otherwise were not able to afford the anti-HIV drugs.

For this trial, Dr. Kabara provided Lauricidin, his patented, 95%-pure monolaurin from coconut lauric acid. Dr. Dayrit, along with Dr. Tayag of the San Lazaro Hospital, an infectious disease center in Manila, screened the HIV-infected patients. Only those who reported to the hospital regularly but had never previous taken anti-HIV medication were included in the study (221).

Among them, fifteen patients were selected, at random, and each was assigned, also at random, to one of three treatment study groups. One group was given coconut oil; the other two groups were given Lauricidin at either a low dose or a high dose schedule. One of those randomized to the high dose Lauricidin group was later found to have a viral load too low to count. This patient was therefore dropped, leaving 14 patients in the study.

The dosage schedules were based on the half coconut consumed daily by the HIV-AIDS patients in Dr. Enig's anecdote. Dr. Dayrit calculated that half a coconut has about 20 to 25 grams of lauric acid, which produces about 7.2 grams of monolaurin. This was the high-dose monolaurin study group, members of which were administered three 800-milligram capsules, three times daily. The patients in the low-dose monolaurin study group were given 2.4 grams, one-third of the high-dose group, administered as one capsule, three times daily. The daily dosage schedule of the coconut oil group was calculated as 50 milliliters or 3.5 tablespoonfuls—about the amount of oil you can get from the meat of half a coconut.

This pilot study went on for six months. By the third month, viral counts had decreased in seven of the fifteen patients, and by the sixth month, the counts had decreased in nine of the fourteen patients. Of these nine patients, two had been on the high daily dose of Lauricidin each day and four had been on the lower daily dose.

Unexpectedly, three members of the study group of five HIV patients who were *given only coconut oil also had a decrease in their viral counts.*

This is an open preliminary study with very few subjects and no controls. Yet, it begs the question—*What reduced the viral counts in three of the five who took just coconut oil?*

Dr. Dayrit explained the effect using a comparison with human breast milk, which contains about 17% lauric acid and is said to protect babies from infection before their own immune cells become active.

Rx: Coconuts! (The Perfect Health Nut)

Do you recall from Chapter 2 that medium-chain and long-chain fatty acids are different? Fatty acids with long chains go across the intestinal wall, where they become recombined with glycerol and are remade back into their original triglyceride form on the other side of the intestinal wall. That was the example about sending chocolates through FedEx.

In either mother's milk or coconut milk, the medium-chain lauric acid breaks down into lauric acid monoglycerides and then into lauric acid. These medium-chain molecules are very small. Unlike the longer-chain fatty acids and their monoglycerides, those of lauric acid are small enough to be directly absorbed through the intestines, without being recombined into triglycerides.

The importance of this distinction between the absorption of the medium-chain and long-chain fatty acids and their monoglycerides is that, as I mentioned above, lauric acid and its monoglyceride are active antiseptics. The triglycerides are not.

The lauric acid/monoglyceride that is absorbed through the intestinal wall can quickly get into the bloodstream and go to your liver, but can also travel around your body to potentially act as an antiseptic. Coconut oil has 50% lauric acid—and another 15% of the shorter C_8 and C_{10}—which have been described to also have some antiseptic properties (222).

Dr. Dayrit therefore reasons that the readily absorbed C_8, C_{10}, and C_{12} monoglycerides from the breakdown of coconut oil in the gut is the reason the coconut oil brought down the HIV viral counts of the patients in their study.

Even in the Asian medical literature, these promising results on HIV caused hardly a ripple, although two or three countries have recently indicated joining up with Dr. Dayrit to replicate and pursue these studies. The incidence of AIDS is quite low in the Philippines. The preliminary plan therefore is for these studies to be done in countries with a much higher incidence of HIV/AIDS.

My own clinical research studies started at about the same time, not because of the HIV study but because I met Dr. Kabara in 1998. I started to become curious about the antiseptic effects of coconut oil, but as a scientist, I needed to do a hands-on study to convince myself that this derivative of a food I enjoy eating could actually kill organisms on my skin. So, I started with two clinical studies on hand bacteria.

Clinical Study Number 2: Dermatological Study of Hand Gels (223). In this study, we chose an FDA-mandated protocol to measure the effectiveness of hand-sanitizer products. This protocol makes use of *Serratia marcescens,* a disease-producing bacterium that usually is acquired from long stays in a hospital. S*erratia marcescens*'s growth in culture plates develops a cranberry red color. This added to the excitement we felt as we made our first observations on monolaurin (yes, we doctors get cheap thrills when our studies feature fun color-changing organisms; chalk it up to long hours in the laboratory). With bated breath, we watched to see if monolaurin could remove that red growth, killing the bacteria.

Since there had been no precedence for the clinical use of monolaurin in studies on the skin, we decided on a small sample of 30 patients. These were healthy hospital personnel volunteers, divided into two groups that applied one or the other of two coded test formulas: 1.5% monolaurin and 70% isopropyl alcohol, both in an identical and bland gel base. We followed the FDA protocol closely for the cultures of the red *Serratia marcescens,* the pretest conditioning, the hand contamination with the organism, and the glove-juice sampling technique for bacteria collection.

The test itself was a three-step procedure, repeated ten times: contaminate the hands with the *Serratia*, wash, and sample, with a five-minute rest after each wash. The samples taken after the first, third, seventh, and tenth washes were cultured.

After a single hand wash, both 70% isopropyl alcohol and 1.5% Lauricidin showed 40% greater degerming activity than the ordinary effect of a regular bar of soap. In addition, both the alcohol and the Lauricidin groups showed a parallel and progressive decrease of the bacterial colony count from the first, to the third, to the seventh, to the tenth cultures.

In other words, 70% alcohol did what it is supposed to do; it removed bacteria. But a much smaller concentration of monolaurin—1.5%—did just as well in destroying those red growths in culture!

Then, when we looked at our volunteers' hands, we noted the Lauricidin users' skin was soft and had an impressive absence of pain or discomfort. In contrast, a significant number—nine of fifteen (60%)—who used the isopropyl alcohol formula reported skin dryness and roughness, with pain or discomfort of the palms. When we examined them, we saw an obvious, pronounced skin swelling, redness, and dryness.

In contrast, Lauricidin demonstrated the same antimicrobial effectiveness as the alcohol-based sanitizer, both immediately and over time, but it did this with the panache of a skin-enhancing or even skin-soothing moisturizer.

We found the results impressive, but as true researchers, we were still skeptical. We immediately planned a study to include not only model pathogens but also real bacteria on the skin of real people who had just finished taking care of patients. We wanted to see what Lauricidin would do to organisms in real time, using actual hospital bugs cultured from after-work hands, not just *Serratia marcescens*.

Clinical Study Number 3: Dermatological Study of Hand Gels (224). Seventy-seven hospital employees, nurses, and caregivers agreed to participate in the next study. They were told to report after duty, without washing their hands. The researchers used the same glove-dip technique, and the sample water was immediately sent to the laboratory to be cultured. The six most prevalent bugs cultured from the employees' hands had the following names: *Staphylococcus epidermidis, Staphylococcus citreus, Acinetobacter baumannii, Pseudomonas aeruginosa, Enterobacter* sp., and the name of a fungus, possibly more familiar to women . . . *Candida*.

This impressive number of organisms is the reason why doctors wash their hands with an antiseptic or put on gloves before they examine you.

In Vitro Part of the Study: In what is called an in vitro study in the laboratory, these organisms were grown in tiny glass dishes with a standard culture chemical, on top of which organisms grow in colonies. A measured portion of the colonies was then exposed to the test gels. This was done with the laboratory technician blind to the products she was examining.

In Vivo Part of the Study: We randomly selected forty-five of the volunteers from whom we had gotten the cultures. We gave them all measured samples of the cultures to apply to their hands in a specified manner. With the bacteria in place, one group of fifteen applied to their hands a saline solution. This was the inactive, control solution. Another group of fifteen applied the Lauricidin hand gel. The third group of fifteen applied the alcohol hand gel. The test materials were given in the same blind fashion; neither doctors nor participants knew what the participants were using. The glove-juice samplings for colony counting were all done as in the previous study, and counts were taken after 30 seconds and after 5 minutes.

In all instances, in vitro and in vivo, the reduction of colony counts was significant and comparable for both the monolaurin and the alcohol groups, and was not significant for the saline control group.

Then, to study the effect of the two products on the skin itself, we taped small test patches with monolaurin gel, alcohol hand gel, and saline solution onto the skin of the backs of each volunteer. These patches were kept in place for 48 hours and then removed; the skin was examined, and then reexamined after another 24 hours.

In nine of the fifteen participants, the skin under the alcohol gel patches was either red or itchy and gave a stinging or burning sensation. The skin under the monolaurin patches produced no reaction in any of the fifteen patients. This test confirmed what we had observed on the skin of the hands from the alcohol applications during the first study.

Clinical Study Number 4: Case Report Number 1 (225). We were ready to use coconut oil, both internally and externally, when a twenty-nine-year-old male with recurring bouts of herpes simplex came to our hospital with a severe flare-up. For eleven years, he steadily had more frequent recurrences, such that in the two years previous to this episode he had experienced severely painful mouth and genital ulcers five or six times each year. With some episodes he had a high fever, which warranted three hospital admissions. At this last admission, he was no longer responding to standard antiviral drugs.

In the hospital, because this patient was no longer responding to the antiviral drugs, we gave this very sick man enzymatic virgin coconut oil to use as a mouthwash and a throat gargle for several minutes after breakfast, lunch, and dinner. He was also asked to drink 1 tablespoon of EVCO after both breakfast and lunch, and 1 1/2 tablespoons for a third dose after dinner. The EVCO was externally applied to his skin lesions, which by now were occurring not just on his genital areas but also on parts of his trunk, arms, and legs.

Our hospital is a well-equipped, modern teaching hospital and this was the first time we had used a natural product for a serious infection. We therefore closely monitored his blood, skin, and general condition.

After five days, the fever came down, the mouth and throat lesions started to become less red and less painful, and the genital and body lesions began to dry up. By the tenth day the body rashes cleared, leaving just dark marks on the skin where the rashes had been. The painful sores and ulcers in the mouth dramatically dried up, softened, and cleared, leaving only trace marks.

Blood tests were all normal, such that he was discharged on the twelfth day. In the subsequent eighteen months, he has been taking 1 tablespoon of coconut oil three times a day, and he remains free of herpes virus sores.

This kind of story, when published, is called a Case Report. Its possible validity is established because we carefully documented daily skin changes, took photographs, and monitored his laboratory studies.

Clinical Study Number 5: Case Report Number 2 (226). Another astonishing case is that of a twenty-year-old male nursing student whom we had to hospitalize one year ago to drain foul-smelling, severely infected scarred skin that was so irregularly thickened it had deep sinuses. He has a skin disease that affects the sweat glands called apocrine. In this condition these glands become unusually active, and exude a lot of smelly secretion. He had these at the neck, the armpits, stomach folds and the groin area, made worse by an inherited trait for developing very thick scars, and obesity. After yet another course of antibiotics, and while he continued to have wet, draining sinuses, we started him daily on 3 1/2 tablespoons of coconut oil, divided into three separate doses, and discharged him from the hospital.

One year after using just this treatment, this young man walked into our office, less obese and with a broad smile, and showed us his skin. Though still thickened by scars and with deep crevices, these were now thoroughly dry (238). All he did for a year was take 2-3 tablespoons of coconut oil each day, as instructed. He did not take any antibiotics.

Since the results of these individual cases are convincing enough to pursue, we plan to observe more cases similar to them. Following that, we will do what are called randomized clinical trials to more fully evaluate the effect of coconut oil as a broad spectrum antiseptic that can be used internally and externally.

Thus, ethically, I must caution you that although I am excited about our findings, these are *preliminary* results that I am sharing with others. *For the treatment of these serious infections, you must follow your own doctor's treatment advice.*

Potentially, yes, coconut oil may eventually help treat such serious conditions on a regular basis. Doctors are always looking for new antibiotics to treat infections that have become resistant to old antibiotics. We may have a need for new ones from the coconut monoglycerides some day. However, for the moment, remember to follow your doctor's guidance for serious conditions such as those described herein.

Finally, let me share with you a beauty secret. Over the past two years, I have asked a cosmetic and skin care laboratory to produce products that are preservative free. As I mentioned previously, one of the main causes of allergic and irritant reactions to beauty products is the common preservatives used to keep products free of microbial contamination. This includes possible entry of organisms from your hands to the makeup you are applying to your face.

Many of the preservatives used in these products are antiseptics related to formaldehyde. Formaldehyde is used to preserve human and animal specimens in the laboratory. These and all other preservatives that release formaldehyde are potential irritants. Many people experience redness, itching, and dryness—an allergic reaction to these preservatives.

For the preservative-free cosmetics I requested, we added a coconut monoglyceride or the Lauricidin as an ingredient to retard microbial growth. Lauricidin is a well-known cosmetic ingredient, prized for its moisturizing effect, a better spread of a product, and a smooth feel on the skin.

Rx: Coconuts! (The Perfect Health Nut)

Just as in the patch testing of the patients in the hand gel and coconut oil studies, monolaurin as a preservative in products was not an irritant and did not cause allergic reactions. Even our patients who are extremely sensitive to cosmetics and skin care products can now enjoy these preservative-free cosmetics. Keep in touch, and I shall let you know once our studies on these monolaurin gel and preservative-free products are finished. Or visit www.vmvhypoallergenics.com to learn about the progress.

Clinical Study Number 6: Dermatological Study of Skin Infections (227). The monolaurin hand gel studies examined bacteria picked up in the course of a hospital day by nurses who were examined and not found to be infected by the microbes. We next wondered about organisms from already-infected skin, which we can actually culture and identify in the laboratory.

This led to still another study. We found the following organisms at the top of the list of common superficial skin infections seen at our hospital's Out-Patient Department: *Staphylococcus aureus, Coagulase (-) Staphylococcus, Streptococcus* spp., *Enterobacter* spp., *Enterococcus* spp., and *Escherichia vulneris*.

After identifying these organisms, we let them grow in nutrient material in small lab dishes. Sections of the nutrient growth material were exposed to monolaurin or to one of six antibiotics that we commonly use for skin infections. These were trusted antibiotics whose use is frequent and widespread. You may even be familiar with their names: penicillin, oxacillin, erythromycin, fusidic acid, mupirocin, and vancomycin.

The study compared how well these antibiotics and monolaurin got rid of—we call it "cleared"—the infectious bacteria in the culture dish.

Are you ready for the results? They surprised us. Monolaurin, across the board, cleared all six of the top bacteria that we cultured from the superficial skin infections. It was comparable or even higher than the clearance by routinely used antibiotics. The antibiotics, surprisingly, were not as consistent as the monolaurin; some had high but others had low clearances, whereas the monolaurin had high clearance for all bacteria.

These results also surprised and impressed the judges at our hospital's

Annual Resident's Research Contest in 2004. The judges scrutinized the method used in the study and nitpicked the protocol to pieces. In the end, they agreed the paper was excellent, and awarded it *first prize*.

How Do Coconut Monolaurin and Monoglycerides work?

Most antibiotics act mainly by interfering with enzyme processes within the bacteria. In contrast, the action of monolaurin is a direct physical insertion into, and breakup of, the lipid coats of susceptible "bad" organisms (228, 229). This lipid coat is a fatty envelope that surrounds the cell membranes of certain bacteria, similar to a shrink-wrap, though thicker. Monolaurin manages to break up the "shrink-wrap" of the bad cells so that they dry up, wither and die.

Other studies show that monolaurin interferes with cell signaling (230-231) and even with the aging of viral cells (232). Still others show monolaurin helps improve immunity against these organisms (233-234). It might be worthwhile to note here that another study even showed that coconut oil protected laboratory rats from the potent shock-type effects of a chemical endotoxin produced by a bacterium called *E. coli* (more below on coconut oil) (235).

The cause of these monolaurin effects still needs further study, just as we need more studies to explain its effect on inhibiting drug resistance, such as that shown by the popularly used antibiotics on the common organisms in the first prize-winning paper discussed earlier (227). In the case of common antibiotics, with frequent use, the organisms develop protective enzymes to block the enzymatic action of antibiotics. Studies need to be done on whether organisms can learn to also block the physical action of the monolaurin on their lipid coat, likewise inducing a resistance to its antiseptic action.

Another important question to answer is regarding coconut oil. How sure are we that enough of these medium-chain fatty acids in coconut oil become converted into the monoglycerides, then absorbed into the blood, to produce anti-infective effects?

We have just begun to address this question with a study that measures the amount of medium-chain lauric monoglyceride in the blood 1, 2,4,6,8 hours after patients drink coconut oil (237). The results of this study, and future pharmaceutical studies, should answer these questions: In the patient studies I have presented, why did taking of coconut oil by mouth have an anti-infective benefit? How much of the oil should be ingested to prevent infection or even to treat it? Does the high coconut oil consumption in the Philippines have anything to do with the country's low incidence rate of HIV/AIDS?

So far we have not seen side effects from the intake of coconut oil. Perhaps a laxative effect, but otherwise we have not seen drug reactions similar to those from antibiotics. Has anyone?

The answers to these questions could prove very helpful for the many underprivileged tropical peoples who may be able to use the coconut not only as a health-giving food but also to affordably prevent or even treat infections. You may have heard, for example, how difficult HIV prevention and treatment is in Africa, where it is pandemic, because of the expense of HIV control. Imagine if the coconut did indeed prove to have HIV-inhibiting properties. Many more studies still need to be done, of course—with more subjects over several years. But the potential could be huge. Millions more poor Africans could begin at least some form of therapy because of the coconut's low cost; they could even make the oil themselves. Moreover, those in more affluent countries who develop lowered resistance to infections may benefit from this food-antibiotic (238).

Initial Conclusions from These Clinical Studies. The information we have so far is this: results from my own studies and those of Dr. Dayrit reported patients with infections that responded to coconut oil taken internally or applied externally. Dr. Dayrit believes monolaurin from the oil's lauric acid is responsible for this effect. We also believe that there are more monoglycerides from the other medium-chain fatty acids in coconut oil that are anti-infective.

Specifically, coconut oil taken internally was shown by Dr. Dayrit's study to reduce the HIV virus that causes AIDS. Monolaurin in a gel, applied externally, was shown in two studies to be comparable to alcohol. Like alcohol, monolaurin can kill common disease-causing bacteria and fungi cultured from the hands of nurses after duty. Unlike alcohol, however, monolaurin is kind to the skin. In yet another study, monolaurin was shown to be as effective as the most popular antibiotics in killing bacteria from already-infected skin. In one case study it has also been shown to be effective as an antiviral therapy.

As far as side effects, monolaurin apparently does not kill off the desirable intestinal bacteria but only acts on potentially disease-forming microorganisms. Currently, we do not know why. These questions and many more about the clinical uses of monolaurin and the coconut oil monoglycerides need further study.

Pending the results of more studies, I do not currently give coconut oil as an antibiotic for infections, but I use it more as a prophylactic antiseptic (meaning to prevent infections more than to treat them). As a medical doctor, I am comfortable doing this because of these facts: First, in the adult HIV patients of Dr. Dayrit's study, 50 milliliters (about 3 1/2 tablespoons) reduced HIV counts. Second, babies receive about 17% lauric acid in their mothers' milk. This lauric acid in mother's milk is believed to help protect the babies from infections, and can thus be considered a prophylactic dose. Third, the herpes virus case report. Fourth, the hidradenitis suppurativa (the apocrine glandular disease) case report. Fifth, from my clinical practice, the still to be reported often secondarily infected psoriasis and atopic dermatitis cases whose severity have distinctly decreased with the regular use of the oil as a nutritional supplement.

Using these facts and numbers, then, I generally tell my patients, *One tablespoon of virgin coconut oil a day to keep the doctor away. Take two tablespoons of virgin coconut oil when you feel run down. When you feel like three tablespoons a day, it's time to call your doctor.*

In a Nutshell

In this chapter, through several clinical studies and case studies, we learned these facts about coconut monoglycerides . . .

➢ Monolaurin, the monoglyceride from coconut's lauric acid, has been shown in the laboratory to have a *broad-spectrum antiseptic* action against bacteria, viruses, and fungal organisms.

- Modern, evidence-based clinical studies show *monolaurin* produced from coconut lauric acid is effective

    o against organisms on the post-duty hands of nurses, and even from already-infected skin; and
    o in decreasing HIV counts.

- *Coconut oil*

    o decreased HIV counts;
    o treated and is preventing the recurrence of a drug-resistant herpes simplex; and
    o treated a case of *hidradenitis suppurativa*.

- Monoglycerides *from enzyme-derived virgin coconut oil* also show promising results against microorganisms.

So far, we've seen the coconut's benefits for your diet and for preventing and treating various infections (on the skin and internally). Having taken care of your insides, let's move slowly outward and see how this Perfect Health Nut can help make you look as fabulous as you feel!

Chapter Six

Far From Giving You Acne, a Coconut Oil Derivative Can Help Treat It

In This Chapter

- The Different Types of Acne
- Coconut Monolaurin Studies: On Treating Regular Types of Acne
- Coconut Monolaurin Studies: On Treating Sweat Acne
- In a Nutshell

Perhaps an instinctive and common association is oil = acne. This is certainly understandable as acne is often caused by too much of the skin's oil clogging a pore. When I tell you that—far from adding to your skin's acne—a coconut oil derivative can in fact help treat and prevent acne, it may sound like I'm asking you to go against all your instincts and to take a leap of faith. But I actually don't need to do that; the studies speak for themselves! Read on to learn more about how a product from this unique oil can actually help control the skin's oil and can effectively treat acne while simultaneously delivering disinfecting and antiseptic actions.

The Different Types of Acne

Maybe you know someone with a similar story, or this story may even apply to you . . . Jane is unhappy, desperate to look good. She went through her teens with hardly any pimples. Now, fresh out of college and putting on makeup every day to look her best for her new job, she is getting pimples. At twenty-three, she has the same ugly red acne bumps that her sixteen-year-old sister is experiencing.

Eager to start her new life looking as good as possible, she had started a serious weight-loss program to shed accumulated fat from too much junk food and too many late-night snack binges. She lifts weights at the gym three times a week and finds time for jogging or swimming. Her gym instructor gave her high-potency vitamins, which she also takes regularly. She even bought a new toothpaste that promised to whiten her teeth so she could beam at her new boss and co-workers with pride. She feels good, looks trim, and is energized, but her new pimples are frustrating all her good efforts and now her back is also breaking out with acne bumps.

ACNE: 101

Acne Types (239, 240). In Jane's short story, I've just related to you five kinds of pimples—or as we dermatologists prefer to call them, acne and acne-like conditions. Jane's younger sister has *acne vulgaris,* the

Rx: Coconuts! (The Perfect Health Nut)

"real" acne common to teenagers when they react to hormonal changes during puberty. The red acne bumps on Jane's face occur when pores become clogged by makeup. We call this kind *comedone, cosmetic,* or *oil acne.* The acne on Jane's back is sometimes called *sweat acne.* This is an acne look-alike, bumps *caused by a fungus.* Additionally, Jane's vitamins may be contributing to her acne—an effect of halogens, the iodides in kelp (seaweed) that are often added to multivitamins and that tend to stimulate acne breakouts, a condition we call *acneiform drug eruption.* Jane's toothpaste, if it contains fluoride (another halogen), could also be causing red bumps around her mouth and chin, called *perioral acne.*

Teenage Acne and Propionbacterium, the Bacteria in Acne. If you had acne in your teens, this is how it most probably developed. Clusters of specialized cells in your body form glands, which produce active chemicals called hormones. As you enter your teens, the activity of the sex hormones begins to really increase. Besides the easy-to-notice changes in your body shape or the timbre of your voice, these hormones stimulate the pores and oil glands of tiny hair follicles in your face. As pore openings thicken, oil fills and clogs up the hair follicles, forming those miserable-looking—though not yet inflamed—squeezable (but don't you dare) whiteheads and blackheads.

We dermatologists give them the formal name comedones. The red, inflamed bumps of acne vulgaris often are directly caused *not* by eating forbidden foods (like chocolates), but rather, by microorganisms normally present but usually on good behavior within your hair follicles. Let me introduce you to the propionibacterium, one of the two kinds of microbes in acne.

Propionibacteria are normally present deep in the hair follicles. As your hormones begin the process of acne, they also grow in numbers. They irritate your hair follicles and activate immune reactions. These result in the secretion of chemicals that in turn excite cells to produce still more pro-inflammatory chemicals. This series of events induces your hair follicles to become red, swollen, and filled with pus. More often than not, they end up producing ugly, angry-looking red marks or even pocklike scars.

These are the acne we dermatologists call *cysts;* you call them *zits* (or, depending on the cyst's size and my patient's age and desperation level, evil-planet-Mars mega-zits).

Since cysts are caused by a bacterium, your doctor frequently prescribes antibiotics as part of their treatment. Clindamycin and erythromycin are two popular oral antibiotics that are often also applied on the skin. Azelaic acid is a prescription-type topical (meaning *on the skin*) antibiotic. More popular is benzoyl peroxide, which often is available as an over-the-counter gel, lotion, or toner. These work to clear cysts, but you can become resistant to them, or—worse—they may even give you rashes (and a rash can mean either having to stop treatment or risking infections that can further clog the pores). Having another medication available is always good, to broaden your treatment options.

Monolaurin Gel Worked for the Hand Bacteria . . . Can it Work on the Acne-causing Bacteria? After seeing the impressive results of the monolaurin studies on hand bacteria, this is the question my dermatology residents and I asked ourselves. We had just finished another laboratory study that showed monolaurin in as low as 1.5% dilution to be as effective as, or even better than, other commonly used dermatological antiseptics against several kinds of bacteria. It was time to find out if we could make use of monolaurin's antiseptic action against the follicle bacteria, which produces the inflamed cysts or zits of acne vulgaris.

Coconut Monolaurin Studies: on Treating Regular Types of Acne

As we've done throughout this book, let's look at the scientific facts and clinical studies that support the coconut's potential as a treatment for acne.

Acne Study Number 1: On Regular Acne (a) (241). Patients with mild and moderate acne vulgaris, seen at our hospital's outpatient clinic, were included in two continuing studies. The first study was a preliminary open trial without placebo/control, to try to first look for any indication of monolaurin's ability to clear acne.

After the participants washed their faces with a bland soap, they were told to apply to their whole face, twice a day, a 2% monolaurin gel preparation. This was slightly stronger than the 1.5% hand gel we used in our previous two studies. We followed up with the participants every two weeks for a total period of six weeks. We wanted to see the effect of the gel on their acne and to observe any untoward skin reactions.

At the start of the study, which we call the baseline, and at each follow-up visit, we counted the acne bumps that were red and elevated, called *papules*, and the bigger ones with pus, called *pustules*. The numbers of papules and pustules showed a decreasing trend, seen from baseline to Week 2, Week 4, and Week 6. Statistical computation revealed these differences to be significant, which means, meaningful.

We calculated the mean of the pimple counts at baseline. Then, we compared the means at baseline to the means at the end of the study. A decrease of more than 50% was considered a treatment success; less than 50% was considered a treatment failure. *Treatment success,* with a 58.8% mean lesion count decrease, was seen in ten of the seventeen patients. The other seven were *treatment failures,* with a 45% mean lesion count decrease.

There was no significant irritation from the use of the 2% monolaurin gel on the skin of these participants.

We considered the various possible reasons for the pimple-count decreases with the use of monolaurin. Was it the 40% alcohol, used to dissolve the waxy monolaurin? We doubted this hypothesis because this concentration of alcohol had been shown not to have any antiseptic action. Neither does 2% have an oil-dissolving effect similar to benzoyl peroxide, a common anti-acne product.

We therefore concluded that the decrease in the lesion counts must be attributed to the antibacterial property of the monolaurin. In addition, this happened as early as two weeks after beginning use, and this effect occurred without adverse skin reactions.

Acne Study Number 2: On Regular Acne (b) (242).

We compared monolaurin gel to its gel base in a classic randomized, double-blind clinical trial. Twenty-four patients used the monoalurin gel while another twenty used the plain gel vehicle base.

At first, the results of this follow-up study confused us. We had expected the monolaurin gel test product to be superior to its vehicle gel. We were surprised to find that there was no significant statistical difference between the monolaurin gel and its gel base. There was no difference in the mean number of papules and pustules at baseline, and at Weeks 2, 4, and 8. Neither did patients or physicians notice a difference in the patients' acne.

We therefore looked more carefully into the study and realized that we had attributed the successes in the first study to the monolaurin alone. Apparently the gel base itself, which does have drying effects on the skin, was therapeutic! We had assumed that with more patients, we could get a 60% difference between the test product and its gel base (which we had thought inactive). Based on this assumption, we calculated that with just forty-four participants we could show the difference between the two products.

We were wrong, but we learned several lessons from the results of this study. First, the monolaurin gel again showed it could clear acne cysts by 58.8%, about the same as 58.3% in the preliminary study. Second, the gel base alone is also effective, contributing to the clearing of the cysts. Third, when comparing the two treatments, we needed a much bigger sample of participants to show any statistical difference. Finally, there was no peeling, stinging, itching or redness with the monolaurin gel. If you have ever used acne treatments, you're probably familiar with these irritations that often occur before the acne itself goes away.

Our conclusion is that this 2% monolaurin gel preparation shows strong positive indications of its possible effective and safe use in the treatment of acne. We are now pursuing this study on a larger scale, to learn the extent of its usefulness and safety. In the meantime, however, more conclusive results were shown in our studies of monolaurin's use as a possible treatment for sweat acne.

Coconut Monolaurin Studies: On Treating Sweat Acne

As unappetizing as this may sound, yet another microbe lives in happy coexistence with you in your hair

Rx: Coconuts! (The Perfect Health Nut)

follicles, and it's not a bacterium. It's a yeast-type organism called *Pityrosporum* fungus (243-245). These organisms are present on 90% to 100% of the surface of healthy skin, including the hair follicles—especially those on the chest, shoulders, and back. On these areas, and sometimes on the face, they can produce acne vulgaris-like lesions called pityrosporum folliculitis (PF), or sweat acne. They are the cause of Jane's back acne.

Sweat acne is seen at any age. It is more common among the more physically active, since these people obviously sweat more. Although this type of acne has also been called tropical acne, any regular sweaty activity can produce it, even during winter. Other conditions that can cause PF are diabetes or any disease that causes frequent fevers or poor immunity.

The pimples of PF are generally quite easy to spot. You might even have them. Look for uniform-looking tiny, itchy bumps without heads, on the upper back, chest, neck, or arms—or even on your face. Taken as a single spot, these do not have the attention-focusing appearance of zits. But PF spots are bothersome because they often are widely spread out in these affected areas, much like a sweat-induced rash but with pimple-like bumps. Some of these bumps may redden and become as big as zits. Even worse, these pimples leave conspicuous dark marks that are obvious if you have dark or tanned skin.

Since the microbe that thrives in these sweaty follicles is the fungus *Pityrosporum*, antifungals rather than antibiotics are used to treat it. PF can quite often be mistaken for acne vulgaris, but if only antibiotics and not antifungals are given for PF, this type of acne can last for many annoying years, making it impossible (or at least embarrassing) to wear fashionable, skimpy, skin-baring tops or even simple sleeveless shirts.

Common names for the antifungals used against the fungus in sweat acne are miconazole, selenium sulfide, zinc pyrithione, and ketoconazole, all applied as a lotion, a shampoo, or taken as an oral medication. All are effective, but when the treatment is stopped, the PF spots return almost immediately.

Acne Studies Numbers 3 and 4: On Sweat Acne. At my own clinic, I had observed that patients with recurring sweat acne responded well to the 2% monolaurin gel (246). My residents and I therefore proceeded to acne studies 3 and 4 (247, 248), comparing 2% monolaurin and 2% EVCO monoglycerides (both in a similar gel base) with ketoconazole lotion's proven effects on sweat acne. These were randomized clinical studies.

On our patients' backs, we made a 10 x 10 cm grid to mark the boundaries of the test area. This grid was placed between the shoulder blades. In the test area, the acne spots were counted. If the PF acne spots were on the arms, we placed a similar but smaller grid on the arm. These test areas were recorded on a body form sheet.

As in our previous studies, both the participants and the investigators were blind to which product was randomly assigned to them. The participants were told to use a mild, baby-type body soap for bathing, and to apply the product on the test area in the morning and at night.

At baseline and at the weekly follow-up visits, the investigator, besides counting the number of pimples, also did a microscopic examination for the number of pityrosporum spores within them and on the skin surrounding them. These, together with the description of the pimples, were used in grading the severity of the acne.

We judged the response to treatment as *cleared, improved, no change,* or *worsened,* based on a specified percentage reduction of the lesion and spore counts, and the change of severity grading from baseline.

The results of this study were much more clear-cut than in the acne vulgaris study. Of fifty-four patients, nineteen used the 2% monolaurin gel, eighteen used the 2% coconut monoglyceride gel, and seventeen used the 2% ketoconazole. The baseline data for all patients was comparable.

Treatment outcomes were also comparable. For both the 2% monolaurin gel and the 2% ketoconazole, the success rate was 100%, and for the 2% coconut monoglyceride gel, the success rate was 94.4%! The statistical analysis of all these outcomes showed no significant difference among the three test products.

Nor were there significant adverse reactions among the three treatment groups, although all the patients using the gel formulations of monolaurin and coconut monoglyceride said that they liked the cooling effect of the products. Again, we saw that the gels with monolaurin or the coconut monoglyceride are kinder to the skin even as it enacts effective treatment.

Rx: Coconuts! (The Perfect Health Nut)

In a Nutshell

The studies I've shared with you here are the first dermatological studies on the use of coconut monolaurin and coconut monoglyceride's antiseptic effects on *Propionibacteria,* the gram-positive bacteria in facial acne, and on *Pityrosporum,* the fungus of sweat acne. What we have learned is that:

> Monolaurin as a 2% gel appears to be another safe and effective skin treatment for *facial* acne, presumably due to its antibacterial effect on the *Propionibacterium* acne bacteria.
> Monolaurin as a 2% gel may even also be effective for sweat acne folliculitis, which is caused by the fungus *Pityrosporum.*
> Coconut monoglyceride 2% gel appears to also be effective and safe for sweat acne, like monolaurin.
> Possibly because of the gels' moisturizing effect, side effects often seen with other anti-acne preparations were minimal when the gels of monolaurin and monoglyceride were tested.

For people with acne, products from the coconut derivatives are promising, suggesting that by using them, acne-sufferers may enjoy acne-free, clear, and healthy skin. Now what about your search for the perfect moisturizer? Can the coconut help there, too? Very much! Go on to the next chapter to learn how coconut oil is an ideal moisturizer, and then some.

Chapter Seven

Virgin Coconut Oil Is More Than Just an Ideal Skin Moisturizer

In This Chapter

- Why Skin Becomes Dry
- The Cellular Reason for Dry Skin
- Studies on Coconut Oil as a Moisturizer: Its Fatty Acids are Native to Skin
- Potential Antiseptic and Antioxidant Properties
- In a Nutshell

Dry skin is very common. You don't need to be a dermatologist to recognize this skin problem. If your skin looks dry and scaly, feels rough and itchy, you have dry skin. Besides feeling uncomfortable—taut, pulled or even stinging—dry skin can look unattractive. It can emphasize lines and wrinkles, show red blotches, or, in severe cases, exhibit large flakes or scales. While skin that has too much shine is not desirable, neither is dry skin that can look dull, tired or even raw (as it tends to be irritated easily). When we dermatologists think of healthy skin, we imagine skin that is supple, pliant, clear, smooth, and naturally bright and glowing. And, looking at the plethora of cosmetics in the market created to mimic this look—shimmer powders; subtle, "just-pinched" cheek tints; glimmer-highlighting creams—it's clear that this image of healthy skin is very much sought after by the general public, as well.

Why Skin Becomes Dry

You may have noticed that skin has a tendency to change. In hot summer months, for example, normally clear skin can break out in acne. Or, previously unblemished skin can erupt in pimples during hormone-frenzied teenage years. Similarly, normally soft, supple skin can become dry—during certain times of the year or seemingly permanently, over time. Why does this happen?

Year in and year out, skin may become drier as the cooler weather begins. During the fall and winter skin often stays dry and even gets scaly. Outdoors, this is due to low humidity in the air; indoors, it results from dry, heated rooms. In the summer, the air becomes more humid, but the heat of too much sun also can dry out the skin. Or, you may get dry skin from the use of too many soaps, detergents, or antiseptics in the course of your day. Any time of the year, flying long distances or spending a lot of time in an air-conditioned office can have a similar effect.

Dryness of the skin becomes a year-round problem as people grow older, simply because the oil glands of the skin produce less oil. Add years of accumulated exposure to the sun and the problem is even worse.

Rx: Coconuts! (The Perfect Health Nut)

If you have a tendency towards allergies, you may just have naturally drier skin, which compounds the problem even further.

Finally, others have dry skin due to an underlying internal or a primary skin problem. You may have heard of some of these conditions. Some skin diseases are characterized by peeling dry skin. They go by the names psoriasis, or atopic or contact dermatitis. If you have internal problems such as diabetes or kidney or thyroid diseases, your skin can become dry, too. Often this is from the disease itself. But sometimes the skin can become dry from the drugs used to treat the disease, especially if you happen to be allergic to any of them.

Whatever the cause, finding the right kind of oil to apply on skin is so important—for both the feel and look of the skin, as well as for comfort—that the search can become obsessive. Some of my patients will try every new oil they encounter, hoping to find the one perfect oil for their particular needs and preferences. Before I help you in your search, let's first get a more profound understanding of what causes skin dryness. Just like when we learned about fats and oils, this knowledge will help us make more informed decisions about the oils we use on our skin, instead of randomly trying every newfangled oil that comes on the market (which can, if this experimentation leads to an irritation, actually worsen skin dryness).

The Cellular Reasons for Dry Skin

The ultimate cause of dry skin is a defect in the barrier function of skin. This barrier to the environment is a bricks-and-mortar-like one provided by the topmost layer of skin cells, surrounded by a soupy mixture of water and lipids. Different types of moisturizers provide their effect by imitating this barrier.

Instinctively, we treat dry skin by heading for the soothing effects of the water and oil in a moisturizer. An important oil component in many moisturizers is mineral oil. Mineral oil acts as a moisturizer by providing an inert and occlusive film. We call this film *hydrophobic* because it serves to block the exit of water from the skin. By preventing water loss, skin remains soft and pliable. Many dermatologists like using mineral oil because it is available as pure oil, without additives, and rarely if ever produces skin reactions or even clogging of pores.

Due to the positive results so far in studies of coconut-derived products and skin—and again inspired by the local fisherfolk of my home town in the tropics, with their beautifully soft, smooth, unwrinkled skin despite constant sun exposure—I decided to explore virgin coconut oil as a moisturizer.

Studies on Coconut Oil as a Moisturizer: Its Fatty Acids are Native to Skin

My studies had, among other things, revealed to me two facts about coconut oil. First, I learned that in the tropics coconut oil has a long history of use as a moisturizer. Adults and children regularly use the oil to soften, protect, and heal skin and hair dried out by the sun and the sea. Second, after my initial studies showed coconut oil to have potential broad-spectrum antiseptic properties, I realized where I could use this dual effect of a moisturizer that simultaneously removed or prevented skin infections: my own patients with dry skin.

Coconut oil's potential as a moisturizer with broad-spectrum antiseptic properties could help soothe dry skin's itchiness. One reason for the all-too-familiar itchy feeling common to those with dry skin is dry skin's vulnerability to invasion by microorganisms. Sometimes this is caused by a lowered immunity, which in turn is caused by the primary internal problem that gave you the dry skin. Most times, though, the itchiness is caused by breaks that easily occur in dry skin. Think of a dry, brittle leaf, parched and cracked in many places.

Prescribing Virgin Coconut Oil as a Moisturizer. In 1999, slowly at first and then more regularly, I began to use virgin coconut oil in my dermatology practice, especially for patients with psoriasis and for patients who regularly visit our phototherapy day care facility. Then, in 2002, I got a coconut oil that came from the heat-free enzymatic process: enzymatic virgin coconut oil, and its deodorized version. I use them interchangeably with virgin coconut oil from other sources. For more damaged or severely dry skin, however, I like using the heat-free EVCO, feeling it to be a more medical-grade oil. EVCO is also the oil I have been using in clinical tests, rather than just any coconut oil that my patients can buy at the market.

Since 1999, in lieu of mineral oil, I have been prescribing coconut oil to patients with any kind of skin dryness. The results are invariably the same. Patients notice improved appearance, and with it, a reduction of skin roughness or dryness. Coconut oil is easy to prescribe because I find that my patients readily comply with my instructions on how to use the oils. Then they return for refills. This compliance and regularity implies that they are getting good results.

Because of the success I saw in prescribing virgin coconut oil as a moisturizer, I began using it as base oil that I can make more active by adding tars or salicylic acid. This oil softens the scaliness of the scalp, caused by dandruff or psoriasis, and it thins down patches thickened by chronic scratching.

Study on Coconut Oil as a Moisturizer (249). By early 2003, I decided it was time to do a formal study on EVCO as a therapeutic moisturizer, comparing it with mineral oil. To serve as background material for our proposed study, my dermatology resident and I needed to show our hospital's Ethics Committee that coconut oil previously had been studied as a moisturizer. We searched everywhere but found no studies detailed in conventional medical journals. Since complementary—and alternative-medicine research is now being actively pursued at prestigious universities, we looked there, too. We found one study on coconut oil for the hair (250) but could not find a single paper on the subject of the coconut as a skin moisturizer (251).

Therefore, we submitted to our Ethics Committee a broad picture of our use of coconut oil since 1999 and a close scrutiny of forty of these patients. We reviewed the charts and interviewed these patients. The results were similar to the others: they felt good, they liked the oil, they used it, and they experienced no side effects from its use. An occasional patient from our general pool did not like the scent of coconut oil and opted for the deodorized version. The forty we scrutinized, however, all found the somewhat sweet, aromatic smell of the oil on their skin to be pleasant. None had rashes, itching, frank contact dermatitis, or worsening of skin problems from the oil itself.

The ethics committee approved our protocol. We included in the study thirty-four randomly selected patients with mild to moderate skin dryness, particularly of the legs. First, they were patch-tested with the coconut oil or mineral oil test products to ensure that they were not allergic to or irritated by them. This is the same test we did in the hand gel study. None reacted. The patients were then, at random, given either mineral or coconut oil packaged in identical containers, for blinding purposes. Twice a day for two weeks, the patients applied the assigned oil to their legs.

At the beginning of the study (the baseline) and then at each weekly visit, we used special instruments to measure the changes in the dryness of the skin. To show effect, we measured the amount of water and lipids in the skin. To show safety, we measured changes in water loss from the top layer of the skin, and the skin's surface acidity.

Also at the baseline and at each weekly visit, we used a visual scale that ranged from zero to ten and we asked the patient to rate his or her dryness, scaling, roughness, and itching. The investigator separately evaluated these symptoms, using an accepted dermatologic grading of dryness.

By the time we reviewed the results of the study, a large number of my patients and I had become regular and satisfied users of virgin coconut oil. Yet the results of the study still surprised me. Using all the instruments I mentioned, coconut oil and mineral oil significantly moisturized the skin. Clearly illustrated in graphs and tables, the moisturizing effects of virgin coconut oil were even more convincing. And neither oil caused irritation, allergy, or other adverse reactions.

More interesting to me, as a user, was seeing how the test subjects found the oil on their skin. I looked at the subjective grading by the investigators while also looking at how the patients perceived the oil on their skin. The ratings showed that compared to mineral oil, coconut oil had a general trend toward better perceived improvement, better absorption into dry skin, easier spreading, better feel upon and after application, and an acceptable odor.

Coconut Oil's Fatty Acids are Native to Skin. Our study clearly showed that coconut oil has moisturizing effects like mineral oil—but there is a crucial difference between these two oils. Mineral oil is distilled from petroleum, the very same inorganic (nonliving) underground product that is refined to become gasoline for cars. Coconut oil is an organic product that comes from a living plant. *It is made up of fatty acids that are a normal component of the skin barrier,* including those stable, saturated, medium-chain fatty acids—lauric, capric, and caprylic—and the longer-chain myristic acid. All four fatty acids serve as a replacement for lost lipids in the skin's barrier (252).

These same fatty acids, particularly the coconut's lauric acid, with the help of some friendly bacteria, may be broken down into their natural, broad-spectrum monoglyceride antibiotics. We have seen their antiseptic effects on the skin, although the breakdown chemicals from these fatty acids now need to be identified in future studies.

New Studies . . . Increased Potential of Coconut Oil for Dry Skin. These observations gave us new opportunities to study coconut oil. Infections in people with dry skin are often a secondary effect. In turn, the infection serves as a trigger for the primary skin disease to recur. A good example is atopic dermatitis, a form of inherited skin allergy. Children and adults with atopic dermatitis always have dry skin. When this dry skin becomes secondarily infected, the itching and dryness get worse (253).

We are now examining the bacterial and fungal growth in atopic and other kinds of chronically dry skin treated primarily with coconut oil as a moisturizer. An important part of this study is to monitor the effect of the long-term use of coconut oil for its associated antiseptic effects. We want to discover whether the regular use of a moisturizer that also naturally removes organisms can prevent infection, so these diseases will recur less frequently. We also want to see whether these organisms will eventually develop a resistance to the coconut oil's antiseptic action.

Over the past five years, besides watching out for answers to these questions, I also have been alert to adverse skin reactions—as dermatologists always are—when introducing the use of a new product.

Both my patients and I were dubious and curious about the effects of coconut oil on their skin. When they responded well to it, I shared with them my concerns, and together we watched for adverse reactions. So far, our experience is that the virgin and enzymatic coconut oils are virtually innocuous, free of any irritant or contact dermatitis from its use.

Nonetheless, we will continue to monitor coconut oil for its safety, especially now that other doctors also are beginning to use coconut oil in their practices. For instance, some members of our pediatric department quietly observed our studies and listened to my extemporaneous comments on the oil. The advantage of a natural antiseptic from fatty acids that are a part of mother's milk is obvious, especially for premature babies who inherently have very dry skin and are prone to infections. In response, these pediatricians have just started a study to see if coconut oil or its monolaurin can be used for babies with these problems.

Potential Antiseptic and Antioxidant Properties

Skin cancer is rare in the tropics, even among the farmers and fisherpeople who are regularly exposed to the sun but cannot afford synthetic sunscreens. For instance, local island dermatologists rarely see any actinic keratoses, let alone the large numbers that represent the bread-and-butter consultations for my dermatology colleagues in Hawaii or Florida. Naturally, therefore, now that I was on this coconut oil wavelength, I proceeded to study its possible effects on the skin as protective oil.

Sun Damage. First, let's review what happens when we are exposed to ultraviolet light rays from the sun. The primary assault by UV is on colored compounds in the skin called chromophores. Activated by light, these chromophores undergo oxidation reactions with the atoms of adjacent cells. This forms free radicals and more specific fragments called reactive oxygen species, which cause breaks in the DNA in the core nucleus of cells. When this happens, built-in mechanisms and enzymes help repair these DNA breaks, and a brigade of cells appears in the area.

These cells form chemicals to neutralize the reaction and to segregate it, so as to minimize the damage. This is damage control at its best.

While these events take place, blood vessels become dilated and fluids flow into the area, causing redness, swelling, and eventually blistering of the skin. This is acute sunburn. When you get frequent acute sunburn reactions, the cell repair may become less complete and the damage to the DNA becomes more permanent. Dry, discolored, wrinkled skin appears, followed by early and then frank cancer spots.

Sunscreens are the accepted way to prevent these sun-induced changes, which are called *photodamage, photoaging,* and *photo-induced skin cancers* (254). However, a newer and different approach to preventing these changes is now being pursued vigorously by photobiologists. This approach uses antioxidants, because oxidation reaction at the cells is the basic change that occurs (255).

The older, better-known antioxidants are vitamins A, C, E, and beta-carotene (256, 257), but there are now many more. Most are from plants, which have a rich supply of antioxidants because—unlike humans, who can and ought to walk away from the sun—plants cannot avoid sun exposure. Neither can they apply a sunscreen. Thus, nature provides them with built-in antioxidants.

Antioxidants do not have the chemical ability to scatter, block, absorb, or reflect UV light. Sunscreens do this, on the surface of the skin. Unlike the surface-acting sunscreens, antioxidants are expected to be present at the cells, so they can neutralize free radicals as they form. They are more like chemical scavengers, garbage collectors of the free radicals formed by sun damage.

There appear to be some potential winners in this search for the chemical prevention of oxidation damage to cells. These are the chemicals common to green tea and grapeseed, and another one found in soybean extracts.

Plant Antioxidants. Among these plant antioxidants, green tea is one of the best studied (258, 259). These studies were all performed on Caucasian skin. Our own study of green tea (260) and soy isoflavone (261) as a topical antioxidant showed similar results on Asian skin types from the tropics. Sunburn cells, redness, and darkening reactions were all reduced when these antioxidants were applied before exposure to the test light. The studies showed that both green tea and soy isoflavone could prevent darkening of the skin. This is great news for Asian skin types, many of whom prefer *not* to have tanned skin.

I should, however, emphasize and reemphasize at this point: this is excellent news but you must *always* protect your skin with a sunscreen (I recommend both indoors and outdoors), with a minimum SPF of 30+ and PFA (protection factor against UVA) of at least 15. More and more sunscreens probably will also contain antioxidants.

Preliminary Study of Coconut Oil as an Antioxidant. To test coconut oil, we followed the same protocol we used when we are looking at the antioxidant activity of green tea (260) and soybean extracts (261). Like green tea and isoflavone, coconut oil reduced the reddening reaction and the number of sunburn cells. Unlike green tea and isoflavone, coconut oil did not protect against darkening of the skin. This correlates well with coconut oil's perceived usefulness as tanning oil.

It is important to understand that these are exciting preliminary data from only a few participants, which now need to be validated by more studies and more participants. I share these results with you only to let you know about this new breed of antioxidant chemicals, which includes coconut oil. They have the potential to help us protect ourselves from aging, from the skin cancer effects of sun exposure, and possibly from other changes brought on by oxidation in our skin.

Again, I must reiterate: these antioxidants cannot and must not be used in place of sunscreens to protect you from the sun.

In a Nutshell

While the studies regarding coconut oil as a moisturizer for dry skin are still few and far between, coconut oil shows tremendous potential as an effective treatment for dry skin. In these preliminary studies, on multiple patients, it was found that coconut oil

➢ is a good moisturizer;

- is naturally healthy: it is composed of fatty acids that are native to the skin's barrier layer;
- has the potential additional advantage of antiseptic properties; and
- has the potential, based on the results from early preliminary studies, to have a mild antioxidant property that helps protect skin from sunburn, although it appears to darken skin like a tanning oil.

In the next chapters, you'll learn even more about coconut oil's potential for skin and hair—as a growth factor—as well as how to start incorporating the coconut into your diet and personal care regimen.

Chapter Eight

Coconut Water Growth Factors: Can Potentially Retard Aging to Rejuvenate Hair and Skin

In This Chapter

☑ The Discovery of Growth Factors in Coconut Water
☑ Studies of the Growth Factors of Coconut Water
☑ Growth Factors: What Can They Do?
☑ Studies: Kinetin Growth Factor for Anti-aging
☑ The Stem-cell Life Cycle in Hair
☑ Plant (Cytokinin) and Hair (Cytokine) Growth Factors
☑ Future Possibilities for Coconut Growth Factors
☑ In a Nutshell

For us dermatologists, rejuvenation is a field of continuing study. Why? Because while people want us to cure serious skin diseases, they also want us to fulfill their desire to look young. Here are some random examples: first is myself. Now in my sixties, I eat well—coconut products included, obviously. I go to the gym and play tennis regularly. I meditate often and try to think the best of every situation to keep my stress levels low. Through the years, my many patients have become friends, and somehow they forgive me for my frequent travels for work, play, or writing this book. With a hair-raisingly full schedule, why do I make it a point to do all these things? Because by doing them I *feel* young, *think* young, and with some help from my own trusted dermatologist colleagues, *look* pretty good.

I am not alone in wanting to look young at any age. From nineteen to ninety, my patients regularly ask me what's new in the field of rejuvenation. Closer to home, my husband and two daughters will sneak in ambush consultations on something they have just heard or read about. You've probably read of similar "hot" developments in glossy magazines and best-selling books on beauty or watched them on television on some lifestyle channel or even the news.

In these media, the message is often the same: looking young is best. In my office I use various devices—lasers, IPL, radiofrequency, infrared and other light—or heat-generating devices. Botox, and fillers work well together. And, always, broad-spectrum sunscreens with high-value protection against UVA, UVB, and even against visible and infrared lights.

Through the years, my residents and I have investigated products to lighten, peel, and moisturize skin, in toners, lotions and creams. With ongoing studies on the oil and monolaurin from coconut, I all of a sudden thought, What about the water from the coconut?

All I knew about it was that it made the young green coconut heavy with a low-carbohydrate water, containing about 3% natural sugars. Besides loving it as my favorite low-carb drink on a hot summer day, I began to be curious if this part of the coconut, too, had beneficial properties. I knew that, for emergency situations, the Philippines and many other remote tropical countries have reported using coconut water as an "IV" straight from the nut (263). Here is one particularly inspiring

Rx: Coconuts! (The Perfect Health Nut)

story about coconut water. In 1962, practicing in Iloilo, an island in the Philippines, Dr. George Viterbo was given an award for his life-saving efforts using coconut water to assist people struck by cholera, an infection with wrenching diarrhea. Having run out of IV fluids to give, he gave coconut water *intravenously* and saved the lives of many.

So there had to be something more to this seemingly simple juice. Back in the comfort of our shaded garden, I sipped away and I wondered, *What is in this coconut water?*

Among the first things I learned is that like coconut oil, coconut water too has antiseptic properties. That is understandable because of its proximity to the meat that has the lauric acid. Imagine my surprise when I next found out that this water has plenty of growth factors! As you will soon find out, growth factors are useful in the *rejuvenation of cells*.

The Discovery of Growth Factors in Coconut Water

Plant scientists were the first to discover that coconut water contains a group of substances that influence the growth and function of cells. These substances potentially belong to the elite group of contemporary drugs called *biologicals*. Quite simply, biologicals are substances made from biologic or living things. Some common examples of these substances are the products made from blood, and even the blood itself, given in transfusions. The more modern biologicals make use of scientists' improved understanding of what happens at the level of the molecules within cells. By initiating treatments directed at that level, potential problems can be corrected as they arise.

Coconut water has its own biologicals: enzymes, nucleotides, active polypeptides, and growth factors. Just like the antibiotic penicillin, the presence of growth factors in plants was discovered serendipitously from a fungus. In the late 1800s, Japanese farmers commonly observed that their rice grew very tall when infected by a fungal disease called Bakanae disease or "foolish seedling." This fungus was later given the name *Gibberella fujikuroi*. More studies identified the pure compound that promotes this growth as belonging to a group of chemicals called *gibberellins*. Later, more of these gibberellins were found in plants of a higher order than just fungi. They were found in tomatoes, peaches, cottonseeds, guava seeds, and the silk tassel of maize.

Studies on growth factors in coconuts go back to 1941 when the Dutch plant physiologist Johannes Van Oberbeek saw the effect of the coconut's water on the growth of a plant embryo. He mistakenly called this water *coconut milk*, which actually is the white milk squeezed from the meat. However, the naming error is understandable because Van Oberbeek discovered that the plant's baby cells grew faster when coconut water was added to the culture medium (264)—just like mother's milk (or milk in general) helps babies grow. The growth slowed down when the water was removed. Later, it was discovered that this effect was due to the presence of growth hormones in the coconut water (265-267).

Studies of the Growth Factors of Coconut Water

The plant hormones of coconut water are biologically active. Their presence can therefore be detected with tests that make use of their growth-promoting functions. These tests have been done on the cytokinins, gibberellins, and auxins from coconut water. For instance, dwarf rice seeds and their seedlings were observed under controlled laboratory conditions, to discover *when* and *for how long* a sheath of leaf would grow from them. This was done in the presence of coconut water and then compared to the growth without coconut water (268).

These studies have shown that the length of the leaf sheath is directly related to the gibberellin content in the coconut water (268). In these studies, special instruments were used to further identify, separate, purify, and define the gibberellins present in the water. It was discovered that these growth hormones were able to go through membranes and to remain *stable and active* even when exposed to heat, acids, or fermentation.

In 1969, an interesting experiment at the University of the Philippines showed these growth chemicals were hardy, indeed (269). In water from intact ten-month-old coconuts—even when kept in storage—the growth factors continued not only to be produced but also to maintain their growth-promoting activity! The group sampled over a period of 12 weeks, the water from intact coconuts kept in a dark room at ambient 24°C to

Rx: Coconuts! (The Perfect Health Nut)

34°C tropical temperature. Except for during Weeks 5 to 8, the amount of gibberellins increased in the stored coconut water. During the times that these substances were increased, the amount of coconut meat—weeks removed from the tree—also increased (270).

Growth Factors: What Can They Do?

Whether in human beings or in plants, growth factors are produced to regulate growth. These chemicals are very potent. Even tiny amounts produce major growth effects. The plant growth factors have bewitching, Merlin-the-Magician—like names: *gibberellin, auxin,* and *cytokinin.* Each promotes growth. Keeping these in balance are *ethylene* and *abscisic acid,* growth factors with the opposite effect—inhibiting growth.

Gibberellins stimulate cells to divide into more cells and then to become longer. Let us take a dormant seed as an example. The embryo within the seed produces gibberellins, which signal the outer layer of the seed to make and secrete enzymes. Once secreted, these enzymes break down the endosperm within the seed. This allows the embryonic cells to access the seed's nutrients, so the cells can grow and develop. The effects of the gibberellins are enhanced when auxins are also present.

Auxins cause plants to grow longer, to bend toward a light source, and to take root, flower, and bear fruit. Their action is the opposite of the cytokinins, which are produced in actively growing tissues, particularly roots, embryos, and fruits.

Working together with auxins, *cytokinins* stimulate cells to divide, develop, and mature. Cytokinin alone allows the cells to get large, but they do not divide. When cytokinin and auxin are present in equal amounts, the cells divide but do not differentiate; add more cytokinin, shoots develop; subtract cytokinin and roots develop.

Chapter 4 showed the role of signaling in and among cells. Cells are constantly interacting with each other via other cells, enzymes, and hormones, important substances that specifically act on cell differentiation and growth.

This is why the growth factors of plants are also called plant hormones.

Studies: Kinetin Growth Factor for Anti-Aging.

One of the most studied of the cytokinins is *kinetin* (271-274). It was first isolated from the DNA protein in the sperm cells of herring. Because of its effect on cells, kinetin was called a cell division factor. Besides in fish, kinetin was later identified in animals and plants, and found especially abundant in unripe corn seed and in coconut water.

Many of the studies looked for the effects of kinetin on plants when added to their water. What they found was that the plant cells were stimulated to divide, leaves were prevented from becoming yellow, and fruits did not become overripe (275). Plant biologists therefore make use of kinetin to delay the aging and death of plants.

Naturally, the next question for researchers was, would kinetin have the same effect on other living organisms? One fascinating study was done on the effect of kinetin on fruit flies. Kinetin, when fed to these flies, slowed down their development and aging and prolonged their life spans. This was dose-related: the higher the dose, the longer the life span was extended. For male fruit flies, the median life span of thirty-five days increased to forty days; female fruit flies saw an increase to sixty-three days. The oldest male fruit flies had their life span increase from seventy to seventy-nine days and the oldest females' life span increased from seventy-six to ninety-one days (276)! The fruit flies saw an increase in their median life spans by 14% for males and 83% for females . . . imagine adding that many more years to our own lives. Taking U.S. life expectancy tables as a reference, men would see their life expectancy increase from 75 years to 85 years, and women from 80 years to 146 years! Please don't take these numbers too seriously . . . there is a bit of a difference between flies and us.

A continuing study made note of a marked increase (about 55% to 66%) in the activity of *catalase*, an important antioxidant enzyme in these fruit flies (277). How kinetin acts is therefore attributed, in part, to its ability to induce the appearance of more of this potent antioxidant.

The next study was on people. First, human cells were made to grow in culture media inside special laboratory dishes (278). The cells, called *firbroblasts*, are plentiful in human skin. Fibroblasts produce the fibers called collagen and elastin, which give skin its tensile strength and elasticity. As we age, fibroblasts produce less of these fibers causing our skin to lose its bulk ... such that unchecked by current rejuvenating methods, the inevitable happens: eyebrows and cheeks sag, creases deepen, and jowls start to form. Ugh.

The same kinetin that retarded aging in plants and fruit flies, when added to the human fibroblasts in culture, also remarkably retarded the aging of the skin cells. They stayed young looking longer and acted like young fibroblast cells in terms of their growth rate, size, and molecular activity.

The beauty of it all is that these changes happened with the skin cells remaining as normal as can be. They acted and looked young but in a normal way, like other younger cells. No cells behaved independently of the others (a precursor for the cancer anarchy of cells stimulated by carcinogen-like chemicals).

For you and me, the next question is *Can kinetin reverse the aging of our skin?*

Yes. Kinetin, in cosmetics has been tested on people (279). Double-blind trials of .005% kinetin in a skin care lotion, compared with its base without the kinetin, were done for 24 to 48 weeks on sixty-four human volunteers who applied the lotions daily. The skin of the volunteers was examined for the aging changes that are induced by the sun. See if you already have any of these: fine and coarse wrinkles, the early skin cancer spots called actinic keratoses, mottled skin discolorations, dilated tiny vessels, or dryness and roughness that are evidence of water loss from skin barrier breakdown.

Dermatologists from the University of California at Irvine observed the skin of patients in the study. After twenty-four weeks, the subjects' overall assessment of their skin resulted in high marks for kinetin versus the control lotion: for overall results, 60% versus 36%; for skin texture, 80% versus 42%; and for fine wrinkles, 47% versus 36%. The doctors' evaluations were consistent with these observations.

We are currently doing a similar study on .05% and .1% kinetin. The results of these studies, once published, can perhaps pave the way for a better understanding of this very promising product—not just for the cosmetic appearance of our skin but also for the potent antioxidant and age-retarding effects on other cell systems of the body.

Retinoic acid—the gold standard of anti-aging products—produces its effects by encouraging premature cell death, so young cells can develop faster (280). Often, this is associated with skin burning or stinging, or the redness, peeling, and dryness common to anti-aging products. Kinetin does not cause these reactions.

Human growth factors and a chemical called carnosine are also used in anti-aging. Carnosine induces cells to proliferate or grow, but a warning has been made about the potential stimulation of cancer cell development by these chemicals. Kinetin does not have this risk. It has even been safely used with glycolic acid, another anti-aging compound.

How then does kinetin produce its anti-aging effects? Kinetin's exact mechanisms of action are not yet known. The current theories make reference to kinetin's possible effect on the signal transduction in cells and its potential as a powerful natural antioxidant (281, 282). These potent effects have led investigators to suggest that kinetin be considered for other possible applications, not only for delaying aging but also for preventing the degenerative changes that cause pathologic states such as Alzheimer's disease. However, these are questions for others to investigate. Here, we'll continue to discuss kinetin and other growth hormones from the coconut's water and how they affect skin and hair.

Now that we've learned about growth hormones in the coconut's water, permit me the following thoughts about the future possibilities for the uses of coconut water. I mentioned earlier that kinetin induces cultured plant and human cells to divide and that it retards aging. Externally applied on the skin, the cream with kinetin also showed retardation of aging. Before we speculate on similar effects on the growth of repressed or aging hair follicle cells, let's first look at how the normal hair follicle is stimulated to grow.

The Stem-cell Life Cycle in Hair

Stem cells are the cells that have been causing a lot of debate recently regarding the need for them, their uses in research and the ethical implications of their research or use. Their good uses are many. In cancer treatment, stem cells supply the body with fresh, newly minted, noncancerous cells after both good cells and cancer cells are wiped out by chemotherapy. The ethical dilemma about them comes from the fact that these cells are *pleuripotential*. This means they have the potential to become a specific type of cell, or number of cells, or even an organ—and in some distant future, maybe even a clone.

How far away do you have to look for stem cells? Not too far. Scratch your head and you are pointing directly at a whole bunch of your stem cells. The hair follicle is one of a few human tissues that contain stem cells. A reservoir of these stem cells is regularly found just beneath the surface of your scalp in an area that is a part of the hair follicle itself. The stem cells are particularly abundant in an area of the hair follicle called its *bulge* (283).

Let me tell you about the life cycle of each and every strand of hair on your scalp. Like plants, a hair strand has a root part called its *bulb*, which anchors the hair in the scalp. For about two years, the cells in the bulb actively grow the hair, which is why you need a haircut now and then.

After this growth phase, the cells in the bulb are biologically programmed to rest. They become dormant and stop forming hair. The hair that had formed during the growth phase eventually becomes disconnected from the inactive bulb and falls off. That's the one you just brushed off your shoulder.

Then, voilà! After about two to three months of rest, the cells in your hair root-bulb automatically begin to divide again. The hair bulb cells put out a new hair, which grows until it peeps out of your scalp and becomes another hair to add to your crowning glory.

In this cycle of growth and rest, the *reservoir stem cells* move downward from the hair follicle bulge to the bulb of the hair. The reverse happens during the hair follicle's dormant seed phase.

At the hair bulb is an enlarged area, made of the cells that divide to form hair, or to rest. This is called the *hair matrix*, Like a plant, it too has a fertilizer section called the hair *papilla*. This papilla is a spongy area made up of soft tissue, in which are nutrient fluids, blood vessels, and cellular elements that feed your matrix cells. The matrix cells are capable of dividing, producing more cells that move upward within the shell of the follicle to become your hair and its corresponding sheath cells (284).

Plant (Cytokinin) and Hair (Cytokine) Growth Factors

Both plants and hair have a fertilized ground-base area (soil/papilla), a root/matrix system, a stem, which becomes the hair/trunk; and cells that move up and down this trunk. The similarities between plants and hair go far beyond these, to more actively potent ones: their growth factors.

Both plants and hair are controlled by growth factors. The matrix cells in human hair are controlled by growth hormones called *cytokines* (285-291). The plant *cytokinins* are the growth hormones we talked about earlier.

Human Cytokines. One large family of human cytokines is secreted by specific garden-variety, common cells in our bodies. One is called a macrophage. Others are the leucocytes, lymphocytes, and monocytes, which you can read about in your blood report when your doctor orders a complete blood count.

In addition to these cells, those located at the hair follicle itself produce cytokines that control the secretory activity of the dermal papilla. (Remember?—that's the fertilizer section at the bottom of the hair bulb.)

Tissue Growth Factors of Plants (Cytokinins) and Humans (Cytokines). After either plant or human growth factors are produced, some act directly on the cells in the same neighborhood. Others stimulate humoral (that means *fluid*) and cellular immune responses. Still others are proteins that bind directly to receptors on the cell surface. Once there, they direct or signal cells to begin to actively divide and grow into specific parts of the hair. In these various actions, there is always a balance: some promote, while other oppose the action.

Rx: Coconuts! (The Perfect Health Nut)

Human Androgen Hormones. We humans have circulating growth hormones. These are made in distant glands and are meant to affect specific target tissues in your hair. Admittedly, you and I are more complex than plants. Our hairs, too, are more complex; hence this second set of growth hormones, which plants do not have.

The androgen hormones selectively target hairs at the beard area of the face, the pubic area, the chest, and the armpits. Most importantly—since they appear to be an essential part of human vanity—they affect the hairs at the top of our heads.

In the head area, these androgens bind to receptors both in the cytoplasm and in the nuclei of the dermal papilla cells, and in some cells that encase or sheath the hair follicle. This complex of androgen hormone receptors moves to the cell nucleus. This then enables the expression of genes that code and affect the making of the cytokines (292, 293). This means that if you have the genes to be bald, the androgens will see to it that it happens.

And there is a pecking order in place here. If you have the genes for it, the androgens may slowly stop hair from growing at the very top of your head, called the parietal area. The hair follicles in this top area become smaller and smaller, such that the hairs they produce become finer and finer, until they are too tiny to be seen. Yet, for some developmentally selective reason, in the case of hairs lower on the scalp, the released cytokines stimulate the matrix cells to divide into nicely growing hairs.

Since men have more male hormones, men are more likely to get this baldness at the top. Women do, too, but less so.

So, the *cytokines* are the tissue growth factors that we produce at our hair bulb to affect the growing of our hairs. We now know that plant cytokinins, with kinetin as an example, can apparently retard aging—in plants, as well as in humans. Therefore, can the *cytokinins* or growth factors of coconut water perhaps retard aging of hair follicles to help prevent us from becoming bald? To answer this question, let's look some more into the growth factors of the coconut.

Future Possibilities for Coconut Growth Factors

The water which contains these growth factors is readily available: a glassful in a mature coconut. When stored for eight weeks and more, the coconut water becomes an even better source of gibberellin. The studies suggest that this is true also of the auxins and cytokinins, including kinetin. But what is the potential of these growth factors for us?

Coconut Growth Factors: Potential for Skin, Hair, and Health. For plants, the growth factors of coconut water are already available as a natural growth-promoting plant fertilizer. For humans, we know the coconut's water is safe, whether taken internally as a drink or mixed in with the meat to squeeze more milk from the meat. This milk, in turn, is used in cooking. With such a long safety record as food we have started to look at how coconut kinetin of the coconut water can affect cell growth or retardation of aging in humans.

For starters, as I mentioned, we are repeating the study of Dr. Rattan by incorporating kinetin in advanced skin care cosmetics. The results of these studies will be finished this year. We have also started to look at the possible retardation of aging by these tissue growth factors—not only of kinetin but also of a stabilized coconut water applied to the scalp in a hair-growth study.

But that will have to be the subject of another book.

There are so many exciting studies and possibilities for the coconut, coconut oil, its water and meat, and its derivatives . . . . This is why, in Chapter 9 (Rx 1) I urge you to keep an eye out for all the new coconut-based studies and products that are coming. Cytokinins, kinetins, gibberellins, auxins . . . These potions from the bountiful supply of growth factors in the coconut's water could have amazing benefits for our own cell rejuvenation.

Traditionally, tropical island folk have maintained that coconut oil is what makes their hair remain thick, black, and lustrous even in old age. My

Rx: Coconuts! (The Perfect Health Nut)

Filipino grandparents, who lived in a seaside town in the northern tip of the island of Cebu, all had thick, healthy heads of hair until they died—in their nineties. This was long before modern shampoos and conditioners made up of synthetic detergents, perfumes, and silicones came to town.

I remember all those stories about thick black hair, and I have my own memories of hair as black as midnight, with the fragrance of newly made coconut oil. Switch over to me, their very own granddaughter, with gray hairs that I have been coloring since I was in my thirties! I'll be one person excited to find out what these studies reveal about coconut water's potential for hair.

In a Nutshell

This section is in part current, in part futuristic. But then, all research begins this way. A question starts it all. And so far, we have learned the following:

> Coconut water exists for the coconut meat to develop in its germinative and growing months.
> Coconut water contains growth factors, also called hormones, that function much like a culture broth and that get used up in the process of growth of the meat.
> Because of its size, the growth factors in the coconut are abundant compared to other nuts in the plant kingdom. They are kept in a stable environment (the coconut itself), continue to be active, and withstand the rigors of heat and storage—perhaps because throughout its life the coconut is exposed to the sun and the elements at 30 meters or more above the ground.
> Kinetin, one of the coconut's growth factors, has been shown to retard the aging of fruit flies and of human cells in culture, and finally, of people using a cream containing kinetin.
> Much about the hair follicle is similar to the life cycle of plants.
> The tissue factors that affect this life cycle (cytokinins in plants, cytokines in hair) have certain similarities.

> The question to answer with new research is whether one or more of the growth factors in coconut can affect the life cycle of a hair that has become prematurely old.

Since the beginning of this book, you have learned why I have come to call the coconut the Perfect Health Nut. Together we've examined the clinical studies and other scientific evidence that prove its numerous nutritional benefits for weight loss, heart health, and potential in cancer control, by protecting and making cells more stable, therefore perhaps, even preventing cancer. We looked at more studies—how the coconut helps treat acne, works as an excellent moisturizer, functions as a non-skin-irritating but very effective antiseptic, and boosts skin's overall health. While more studies, on a larger scale, still need to be pursued, we've even seen the coconut's potential to treat viral and bacterial infections. In this chapter, we've learned about even more future potentials of the coconut—how its growth factors may help us retard the aging of our skin and our hair cells.

After all this learning, what do we do? Apply it, of course! In Part 2, we go beyond discussions of *why* the coconut is beneficial and look at "prescriptions" of *how* to use it.

Part Two

A DOCTOR'S PRESCRIPTION FOR USING THE COCONUT

Chapter Nine

Rx Number 1:
How to Use Coconut Oil, Water and its Monoglycerides for Internal and External Health

In This Prescription

☑ Virgin Coconut Oil: Massage it on Skin for an Ideal Skin Moisturizer
☑ Virgin Coconut Oil: Drink it as a Health Supplement for Inside-Out Health and to Prevent Infections . . . No Kidding!
☑ Anecdotes: Actual Experiences of Successfully Treated, Coconut-happy People
☑ Monolaurin Gel: To Prevent Acne, as an Antiseptic, and in Preservative-free Cosmetics
☑ Coconut Water: Low-calorie, High-potassium, Low-acidity Health Drink
☑ Watch for New Products from the Coconut
☑ In a Nutshell

Ancient Ayurvedic physicians practiced care that we now call *holistic*. Their kind of medicine was focused on keeping a person's physical and spiritual state in balance to promote overall health and well-being. To them, coconut oil, together with its meat and its milk, was particularly special. They used coconut oil to anoint people in sacred ceremonies, then gave them the meat and water from the coconut for sustenance. Even today, many island cultures feature the coconut in various rites and ceremonies, often blessing the coconut before they eat it, and applying the oil on their skin as a salve or sacred oil.

It is clear that the coconut has a long history of use by various peoples. And now that you've learned about why the coconut is so good, and have seen how to use it in your diet, I'll show you how you can incorporate it into other areas of your life—applied on your skin and taken internally as a health supplement—to achieve your own balance and reap the head-to-toe rewards of this Perfect Health Nut.

Virgin Coconut Oil: Massage it on Skin for an Ideal Skin Moisturizer

The first step here is to learn how to choose the right oil, then I can tell you how to apply it on your skin.

Choose the Oil. Choosing the right coconut oil is rather simple. Just make sure it's virgin coconut oil, which it should say clearly on the label. If you'd like more information on the differences between the kinds of coconut oil currently for sale in the market, check out Rx 2 (Chapter 10), "How to Select the Right Coconut Oil, or 'The Tale of the Two Virgins, Coco and Olive'." The rest of the process of selection is more a matter of personal preference.

So get started! Go out and buy two or more bottles of virgin coconut oil with different brand names—so you can compare which you prefer over time. For now, Asian stores are perhaps the best places you can easily find this oil. At your usual supermarket, try to find virgin coconut oil at the international, ethnic, organic, or other specialty sections. If you are familiar with shopping online, search for "coconut oil" and you'll find numerous sellers of the virgin oil. Or look at the product guide at the end of this book for the web site addresses of some sources that I've selected based on knowing the quality of the products myself or having read reliable reviews about the sellers.

This is probably your first time to pick up a virgin coconut oil, so use all your senses when you observe your new purchase. Its appearance is probably uniquely different from any other oil you've ever seen. The oil changes instantaneously from a solid to a liquid form, depending on the temperature of the room it is in (later, you'll learn how to use it in its different states).

Next, observe the distinctive smell of the coconut. Smell its bouquet, like you would a fine wine. Open another bottle or two and compare the scent of the oils. Apply the oils to different parts of your skin, then select the oil whose aroma you prefer. This is a natural product, subject to the many moods of nature and the way the virgin coconut oil was extracted from the meat of the coconut. For your skin at night, or even after a morning bath, you might learn to prefer the oil with the milder or the stronger coconut scent. This is a matter of personal preference. I myself *love* the stronger coconut scent, reminiscent of lazy resort trips, relaxing spa days, and beach fun (it's like wearing my own private getaway on my skin!).

In a warm room, or after you've made the coco butter into a liquid, the virgin coconut oil looks more like water than like an oil. Put it up against the light and see how clear and pure it looks. In an air-conditioned room, or if you live in a cool country, the oil becomes a firm, unblemished, pure white butter. If you've purchased a bottle of the oil and want to use it in its liquid state but it has hardened from the cold, simply place the bottle under warm running water, then watch it instantaneously become a clear liquid. What I like to do is, immediately after purchasing my bottle of coconut oil, to pour about a third of the oil into a jar with a wide mouth. That way, if I want like to use the liquid oil, I do the trick with the warm running water described above, or I can scoop out spoonfuls of the rich coco butter as-is.

A quick tip for travel (I never go on any trips without my coconut oil!): I suggest taking the bottle with you, not the jar, as they tend to leak.

The best bottle for traveling that I've found is a water bottle or similar that seals snug and tight. I'd also suggest wrapping the bottle in a plastic bag, just in case—with all the jiggling around and baggage handling, spillage can still occur, and the plastic will protect your clothes.

Apply the Oil. Now you're ready to apply to your skin the oil you have selected. This is best done after a wash or a shower before bedtime or in the morning. Recognizing that the natural barrier of your skin is an oil-and-water mixture, leave some water on your skin. Just use a light towel touch to sponge water off you. Then put the oil on your skin.

The amount of oil you apply depends on you, but as a general guide, get at least half a palmful of the virgin coconut oil, apply it to a section of your skin, and massage the oil in. On my dry skin, I pour a generous amount of the liquid virgin coconut oil and then massage the oil in thoroughly. If you are in a cold room or prefer the texture of the butter, scoop out the virgin coconut oil coco butter with your hands and follow the same procedure as above. Note that as you apply the white butter, the warmth of your skin immediately melts the virgin coconut oil back into its more liquid state.

Your work or other people's needs often dictate that you hurry from one activity to another, but at least for these few seconds, I suggest you relish your coconut moment. Claim this personal space and time for the simple enjoyment of your unhurried application and massage of the virgin coconut oil.

As you massage the oil into your skin, watch how the scent develops from the warmth of your body. Aromatherapists ascribe healing effects to the scents of essential oils. Perhaps . . . but I'll leave that up to you to decide for yourself.

Within a few minutes of applying and massaging the oil into your skin, observe the feel, texture, and amount of the oil left on the surface of your skin. Usually, parched skin of the legs, arms, and body will readily absorb the virgin coconut oil—no grease spots will appear on your clothing! Why is this? Unlike mineral oil or petroleum jelly, which are inorganic and remain greasy and hot on your skin, coconut oil has fatty acids similar to those native to the skin.

Rx: Coconuts! (The Perfect Health Nut)

If your skin is less dry, allow a few minutes before putting on your clothes. Wear old T-shirts, until you learn how fast the oil is absorbed by your skin. The coconut-based recipes you will be cooking will make you feel lighter and will make you want to use your best outfits again. But to be safe, do the T-shirt test first.

By the time you go to bed or start out for work, your skin will feel soft and silky and will have just a light, tropical scent that makes it seem just about good enough to eat.

The next step is the simplest of all: be natural, free, and comfortable, because your skin is enjoying the soothing, moisturizing benefits of coconut oil.

Virgin Coconut Oil: Drink it as a Health Supplement for Inside-Out Health and to Prevent Infections . . . No Kidding!

To keep the doctor away, drink a tablespoon or two of virgin coconut oil each day. I placed this here rather than in the food section of this book because I believe that by avoiding infection, you become a healthier specimen, feel good, are more vibrant and energetic, walk with pep in your step, and thus generally *look* good. Infections also sap your energy and glow. When they happen on the skin, they can leave you with marks or scars that are hard to get rid off and that can look unattractive.

This Rx step is quite simple: drink a tablespoon of virgin coconut oil every day. Some people find this easy . . . others learn to tolerate it . . . and the very rare individual can't take it easily. Before you make up your mind, however, do try it as the rewards are definitely worth it. Let me just remind you about what taking one tablespoon a day will do for you.

> Virgin coconut oil has a mild laxative effect. Drinking it can help with constipation problems by increasing regularity.
> Stress has now been shown to shorten your *telomeres* and to hasten aging (telomeres are the ends of your chromosomes that shorten as you age; when they are very damaged, they can cause the chromosome to start to unravel). How can stress affect the

153

telomeres? Stress is often psychological, but infections, too, can produce stress, the physical type. This can hasten aging.

➤ Virgin coconut oil has been shown to contain lauric acid, which is surprisingly similar in makeup to mother's milk. And like mother's milk for babies, virgin coconut oil can help you get rid of potentially disease-causing organisms that you may have picked up from other people. It may even keep at bay the bugs and viruses that are staying dormant inside you, just waiting for your immunity to drop. Drinking virgin coconut oil as a nutritional supplement helps keep the stress of a possible infection away.

Take the virgin coconut oil any way you can. Most of my family, friends, and patients simply pour some onto a tablespoon or two and swallow. Others add honey to the oil and it goes down easily. Still others add the oil to their cereal, use it in their salads, or drizzle it on top of a dessert such as a dish of ice cream. I know someone who uses it to butter her toast! She stores it in the refrigerator, pats some on her breakfast toast, and alters the coconut taste by adding flavors that she likes, such as cinnamon or vanilla powder.

Anecdotes: Actual Experiences of Successfully Treated, Coconut-happy People

Throughout this book, I've kept the information scientific and have given you data from actual medical studies with proven, statistical validity. As a doctor, this is truly the best way to validate any claims. Here, I'd also like to share with you some of the experiences I've had with coconut oil—both for myself and with my patients. In addition to medical studies, after all, it is lived experience—actual results that we see and feel—that we can relate to.

My Virgin Coconut Oil Keeps Away Infections. I keep a bottle of virgin coconut oil in my bathroom, and when I start losing my voice or begin to have a sore throat, I take a swig, and then I gargle with it at least twice a day (my husband does the same thing when his throat acts up). For both of us, it always works! Again, actual medical studies of virgin coconut oil's effect on viral infections or the immune system are still forthcoming, but this has been my experience, and it has been so consistent that I thought it was worthy enough to share.

Virgin Coconut Oil in My Clinical Practice for Psoriasis. I've also seen virgin coconut oil's positive effects in my own clinical practice. I see a lot of psoriasis patients. One patient is a priest who started to have psoriasis when he was assigned to a parish in Paris. There was a pattern to his skin changes. During or soon after a bout of sore throats, fevers, and colds, he would develop the scaly spots of psoriasis, which after a week or so would spread like a conflagration to make his entire body red and swollen. Several times he had to go into the hospital for these recurrences.

Back in the tropics, he continued to have these bouts, for which he received treatment in and out of the hospital. We examined him for organisms, but were never able to culture any microbe. We assumed the organism causing his sore throats was a difficult-to-culture virus.

After still another hospitalization, and fed up with antibiotics, ultraviolet light treatments, and the oral drugs for his psoriasis, the priest readily consented to my suggestion that he start drinking a tablespoon of virgin coconut oil three times a day. He also started to use the oil as a moisturizer for his entire body.

Slowly, we noted changes. He was able to leave for a trip to Europe. Six months later, he was able to go to San Francisco during the wintertime. His colds recurred, but they were not as severe as before. The constant fevers stopped. His by now regularly nasal voice became more normal, and best of all, the psoriasis gradually subsided. He has now been off medicines and clear of psoriasis for at least a year. And he continues drinking the virgin coconut oil, one tablespoon, three times a day.

Another psoriasis patient of mine is a vendor at an outdoor market. Just like many poor people in the Philippines, his living situation is stressful. He lives in a tiny, simple home cramped with his wife and three children. His psoriasis was unrelenting throughout the year, year after year. Invariably, we had to put him in the hospital once a year for severe flare-ups of bright red, thick, and scaly plaques and patches, very high fevers, and the return of a urinary tract infection. After one of these episodes, about 5 years ago, I also put him on the virgin coconut oil—one tablespoon, three times a day. He was thankful to try a new alternative that wasn't as costly as the regular drugs and treatments.

For at least 4 years now, I have seen him only for follow-up, perhaps once a year. At our last psoriasis support group meeting, he stood up to say wonderful things about me, and I felt rather humbled; I felt the words of praise ought to have been for the coconut oil!

Virgin Coconut Oil in My Clinical Practice for Children. I also quite often see children with a skin condition called atopic dermatitis, and I usually treat them with coconut oil, both internally and on their skin. They do much better: they itch much less when they apply virgin coconut oil regularly as their form of moisturizer, and it gives them the benefits of its built-in antiseptic potential. So, not only do the children sleep and look better, but so do the parents!

Again, as a doctor and medical researcher, I must emphasize that the true merit of a medicine, food, nutritional supplement, or skin treatment must be ascertained, proven, and replicated by statistically valid, methodologically controlled, evidence-based studies. However, as a practicing clinician who also teaches, I always encourage my residents to try new and innovative treatments on their patients, particularly those with recurring, very resistant conditions . . . or even on themselves, if applicable. It is this combination of research, studies, and applications in a clinical and lived environment that, to me, truly validates the efficacy of any food or product. Take these anecdotes, therefore, as information complementary to the medically and scientifically proven merits of coconut oil and its derivatives that you've learned about earlier in this book.

Monolaurin Gel: To Prevent Acne, as an Antiseptic, and in Preservative-free Cosmetics

I carry a bottle of monolaurin gel everywhere . . . to apply as a hand gel . . . and when I travel internationally, it's great to apply on the feet after the mandatory removal of shoes at the airport. The wonderful thing about this hand gel is that even though it provides a cooling sensation from its mild (40%) alcohol content, the monolaurin leaves a lovely soft feel to your hands—great for first impressions when shaking hands with people.

As a dermatologist, I treat acne with the range of medicines that is available for us to use on our patients. Beyond those, I also use monolaurin as a preventive gel, especially on those who have naturally sweaty skin or on active people who exercise or pursue rigorous physical activity often.

Besides its effect as an antiseptic, monolaurin gel is great for people who complain of the oiliness of their faces. Applied alone or under a matte type of makeup, it helps control shine for several hours—while still delivering the acne-preventing and antiseptic benefits.

Finally, I mentioned in Chapter 5 that many people are allergic to the preservatives most commonly used in cosmetics. By cosmetics, I refer not only to powders, lipsticks, and foundations but also to skin care products, shaving creams, shampoos, and the like. Luckily enough, there are some truly skin-focused cosmetics companies out there that are beginning to replace traditional preservatives with monolaurin. One company is so convinced of the benefits of monolaurin that it has actually committed to being 100% preservative free by reformulating each of its more than 200 products, using monolaurin, by the end of 2006. Check out the product guide at the end of this book for assistance. And if you happen to be extremely allergic to cosmetic preservatives, do keep an eye open for preservative-free (more exactly, monolaurin-preserved) cosmetics.

Coconut Water: Low-calorie, High-potassium, Low-acidity Health Drink

For us today, this water, especially from the fresh green coconut (or now also available as a prepackaged snack drink) is beneficial because of its low acidity. Soft drinks and other prepared drinks, on the other hand, often are quite acidic, making them potentially irritating to your stomach lining. In addition, coconut water is high in electrolytes (ideal when your potassium is depleted like after exercise). Best of all, coconut water is much lower in sugar and therefore in calories. You may want to consider it as a more natural, preservative-free, dye-free, unprocessed, high-potassium, high-electrolyte, low-acidity, low-calorie, low-sugar alternative to your favorite sports drink! See the comparative nutritional information in Chapters 1 and 12.

Watch for New Products from the Coconut

So much more is left to be studied about the coconut. For instance, those growth factors in the water can be useful not only for giving energy but also for improving physique and stamina. Coconut kinetin is showing promising potential in cosmetics for anti-aging therapies. Be on the lookout . . . new coconut-based products are on their way every day.

In a Nutshell

The benefits of coconuts are clearly far beyond dietary. We've seen that the coconut and coconut-derived products can be incorporated into many areas of our lives, for enhanced overall health from head to toe.

> Used on skin as a daily *moisturizer,* virgin coconut oil helps soften and smooth skin while giving it a healthy gleam and even a mild (natural and nonallergenic) scent reminiscent of seaside getaways.
> Drinking a *tablespoon of coconut oil each day* helps fight infections, helps your skin look good from the inside out, and delivers directly into your system all the benefits that you've read about this oil.
> While the clinical and medical studies are impressive, you've heard some *anecdotal success stories* of coconut-happy people, and examples of how they use coconut oil in their daily lives.

- Monolaurin gel helps *prevent acne* and control shine. It also functions as an excellent, non-skin-irritating *antiseptic* for frequent use to combat all the bugs you pick up on a daily basis. And *monolaurin in cosmetic products* allows those allergic or sensitive to traditional preservatives to use safer makeup and skin care products.
- Coconut water is a *multifaceted health drink:* natural, preservative free, dye free, unprocessed, high in potassium, high in electrolytes, low in acidity, low in calories, low in sugar . . . and incredibly tasty and refreshing!
- This is *just the beginning* . . . You can look forward to oodles more health-giving products from the coconut.

Now that you've learned some of the ways to incorporate this healthy nut into your daily care regimens, let's move on to incorporating the coconut into your diet—by first learning how to select the right coconut oil for your food.

Chapter Ten

**Rx Number 2:
How to Select the Right Coconut Oil,
Or "The Tale of the Two Virgins,
Coco And Olive"**

- ☑ The Tale of the Two Virgin Oils
- ☑ Specifications of Virgin Olive and Virgin Coconut Oils
- ☑ Coconut Oil Is Produced by Simple, Natural Means
- ☑ Different Kinds of Coconut Oil
- ☑ Which Grade of Coconut Oil to Use . . . Superb Simplicity
- ☑ The Ultimate Low-carb, High-fiber Flour . . . Coconut Flour!
- ☑ In a Nutshell

B y now, through our explorations of scientific studies, a little bit of chemistry, and some clinical and personal anecdotes, we've seen the many reasons why the coconut can be considered the Perfect Health Nut. There is another excellent oil worth more in-depth examination at this point, one that has justifiably grown in popular and scientific reputation and use: olive oil. Now that you have more knowledge—about fats and oils in general, about how they affect your heart, health, weight, and cells, and about their potential to prevent cancer—we can have a more enlightening discussion about the differences between these two great oils, coconut oil and olive oil. It may surprise you to learn, for example, that coconut oil actually contains about 6% oleic acid, the main fatty acid present in olive oil! Or that the standards of "virgin" and "extra virgin" used for the more persnickety olive oil do not apply to the fuss-free coconut oil. In this chapter, far from replacing your trusted olive oil, you'll see how you can *expand* your "trusted oils collection" from just olive oil to include tasty, healthy coconut oil. More, not less, variety!

The Tale of the Two Virgin Oils

B oth these oils are as old as time. Archaeological records describe the eating of olives more than 35,000 years ago and the cultivation of olive trees for oil in Palestine and Syria since 6000 BC. Many ancient oil mills still exist, although the first recorded oil mill is dated to 1000 BC in Palestine. The coconut tree is also ancient. Ayurvedic records from about 6,000 years ago describe, often and in great detail, the use of the coconut as an essential part of a well-balanced life—as a health food, as a medicine, and in religious rituals—to complete the totality of a person.

The term *virgin* to describe these two oils is a more modern concept. The formal definition of *virgin* refers to a person who has never had sexual intercourse. For Christians, it is the name reserved for Mary, the mother of Jesus. In ancient Rome the vestal virgins were pledged to remain as such (alas), so as to tend the sacred fire in the sanctuary of Vesta, the goddess of the hearth.

Beyond this original meaning, *virgin* is often used to describe a *thing,* to indicate something without alloy or modification, as in virgin gold; something not previously exploited or used, as in virgin timberlands; something which is first, as implied by the advertisements of Richard Branson's records and airline; or something pure and unsullied, when used to describe coconut and olive oils.

For the olive oil, being given the name *virgin*, and its categorization into three basic types of virgin oils, was standardized on June 6, 1996, at the 74th International Olive Oil Council (IOOC), and is accepted by most international bodies on olive oil. The application of this term to coconut oil is much more recent. The Philippines' Coconut Authority, the Philippine Bureau of Standards, and the U.S. Food and Drug Administration have come up with standards that specify why and how a coconut oil is virgin.

People are more familiar with the specifications for virgin olive oil and tend to conclude that the term *virgin* is exactly the same for both oils. This is far from the truth, and the reason, again, is chemistry. What you'll learn at the end of this chapter is that all virgin coconut oils are essentially equal. Their specifications are the same. They differ in the manner of extraction, but there is no such thing as an "extra virgin" coconut oil. A virgin coconut is a virgin coconut.

The Chemistry of Olive and Coconut Oils. The primary difference between the two oils is in the number of saturated fatty acids. How purity is achieved is also different for both oils and is based on their chemistries, which are different. In Chapter 4, you saw that even a simple thing like the location of the first double bond makes a major difference in how these edible oils produce effects in our cells. In the following table, do you notice the difference between the two oils? Coconut oil's fatty acids are mostly *saturated* (92%), whereas those of olive oil are mostly *unsaturated* (83%).

Basic Differences in the Chemistry of Olive and Coconut Oils

	Virgin Olive Oil	Virgin Coconut Oil
Unsaturated fatty acids	Total: 82%	Total: 8%
Monounsaturated (MUFA): Oleic acid	71%	6%
Polyunsaturated (PUFA): Linol*Eic* acid	10%	2%
Linole*Nic* acid	<1.0%	
Saturated:	Total: 18%	Total: 92%
Free fatty acids (FFAs)	Extra virgin to ordinary olive oil: <1% to 3% Because of unsaturation, avoidance of light, air, and heat is important.	0.5 % maximum Because of saturation, this is easy to achieve, even with heat.
Time of oil extraction after harvest, and why	Within 24-48 hours, to avoid formation of FFAs.	Within 24-48 hours; aflatoxin mold may contaminate the meat.
Moisture content	Not specified	0.1% maximum, important.
Heat during extraction and storage	Must be avoided, otherwise more FFAs form.	Needed to achieve lowest moisture content. Does not affect FFA level.

Recall that the basic unit of either oil is a triglyceride that is made up of glycerol and three fatty acids. Recall, too, that double bonds in the chemical formula of an *unsaturated* fatty acid make the oil unstable. When subjected to *air, heat, and light,* these double bonds readily oxidize, releasing the component fatty acids from the oil's triglycerides. These released free fatty acids (FFAs) and their breakdown products give a pungent, off-flavor odor to the oil that, on frying, makes the oil foam.

This, then, is what makes the two oils different: the coconut's *fewer* unsaturated fatty acids (8%) release *much less* FFA, while the *mostly* unsaturated (83%) fatty acids of olive oil release *much more* FFA.

The difference in the number of fatty acids released during processing dictates the processing of the oil to make it virgin.

Because of the much greater number of unsaturated fatty acids in olive oil, exposure to *air, heat, and light* must be minimal in order for it to be virgin. Care must be taken—from when and how the fruit is picked, to the immediate extraction, and then to the storage of the oil. All these careful steps assure that the virgin olive oil you place in your salad is indeed *virgin*.

To keep the unsaturated fatty acids of olive oil pristine, the olives must be cold-pressed, meaning pressed without the use of solvents, other chemicals, or heat. A mechanical press called an expeller is used, followed by water as a wash, a centrifuge to spin off excess water, and a filter to remove sediment.

Note that the FFA content standard for coconut oil is 0.5%—half the standard for extra virgin olive oil, which is 1% or less FFA. Coconut oil easily achieves this standard without heat and even with heat to remove moisture or to make the coconut milk thicker. Extra virgin olive oil has to be *cold-pressed* in order to achieve its standard of 1% or less.

The low FFA content of virgin coconut oil is a given, simply because it is mostly saturated. Fortunately, cold-pressing is not needed to make low-FFA coconut oil. Hence, while for virgin olive oils, being cold-pressed and heat-free are important, heating *is* needed to ensure that coconut oil remains virgin. Dry-press and cold-press plus heat both are fine for the coconut. Coconut oil's FFA will naturally remain low, easily.

Another Difference in Chemistry: The Moisture Content of Virgin Coconut Oil Needs to Be Low for it to Remain Virgin. The more important characteristic for the virgin status of coconut oil is its water or moisture content. A fully mature coconut is uniquely different from olive fruits. The coconut, which is harvested at about nine to ten months, contains about 250 grams (about a glassful) of water. Bathed in this water as it thickens, the meat, which contains the oil, also contains at least 50% moisture. In extracting the oil from the meat, as much of this water as possible should be removed, because water is a good medium for microorganisms to grow. More water than the standard of 0.1% may encourage organisms to grow and can make the virgin coconut oil rancid.

Although many more antioxidants have been identified in virgin olive oil, both it and virgin coconut oil have heat-sensitive Vitamin E. No studies have been done to identify other antioxidants in virgin coconut oil; however, as you saw in Chapter 7, when applied on the skin, virgin coconut oil showed antioxidant effects.

Ensuring the stability, the virginity, of coconut oil is important. Heat needs to be used to reach the standard moisture content of 0.1%. Vitamin E may be lost as a result of this use of heat, but it is easy to supplement. It is a small price to pay, compensated by the many other health benefits from virgin coconut oil. The removal of the vitamin E does not, for instance, reduce the effect of virgin coconut oil in its protection of cells, and possibly against cancer.

Specifications of Virgin Olive and Virgin Coconut Oils

The care given during the process of extraction produces three kinds of virgin olive oil: extra virgin, fine virgin, and ordinary. The main difference is based on that free fatty acid content we talked about—less than 1% for the extra virgin, a maximum of 1.5% for the fine virgin, and a maximum of 3.3% for the ordinary.

Virgin coconut oils all have similar specifications (see the comparative table later on in this Chapter). Therefore, they can *all* be used for any of your cooking needs. The only time not to use a coconut oil would be in the presence of an "off" smell, which means the moisture content of 0.12% or less was not achieved or that moisture somehow got into the oil after the container was closed. Take note of that, do not use that oil and buy another brand next time.

Coconut Oil Is Produced by Simple, Natural Means

Having learned that it took at least sixty years before trans fats were pronounced to be bad for you, you may still have reservations about how coconut oil is processed. You may wonder if, in the making of this oil, there is anything in it as unhealthy as trans fats, such as what happens in the partial hydrogenation of the PUFAs. A simple answer is that coconut oil has been used quite safely for thousands of years as food (189). Validation over this length of time ought to be good enough. But, as our goal throughout this journey has been to rely on scientific, modern information, we shall continue along that road, to prove that coconut oil is produced in a safe, simple, natural, and healthy manner.

The Methods of Extracting Coconut Oil from Coconut Meat: Wet-press + Low Heat = *Regular Virgin Coconut Oil.* This is the traditional wet process and the method still used today by village farmers and housewives in the tropics. The coconut meat is cold-pressed, then exposed to low heat. In this process, the wet meat is removed from the nut with the use of a grater. Then, to squeeze the oil out of the grated meat, farmers use bare hands or wrap the meat in a fine muslin cloth. In small cottage industries, the farmers use small hydraulic presses. This releases the coconut milk, a pure white mixture of coconut water and its oil. It looks very similar to the cow's milk that you are more familiar with. More water is added to the coconut meat, to keep squeezing out more virgin coconut milk.

Coconut milk has many uses, besides being the source of the virgin coconut oil. It is also directly used in preparing food or is packed fresh-frozen, in UHT boxes, or canned. This you can easily find in certain supermarkets.

Virgin coconut oil from coconut milk is made by separating the oil from the water in the milk. In part, this is a simple physical process, based on the difference in density of water which is heavier, and oil which is lighter. When the milk is allowed to stand in a covered vessel the (heavier) water settles in the bottom of the vessel while the (lighter) oil naturally separates from the water and rises upward. Left undisturbed longer or overnight, a natural process called fermentation occurs, Fermentation is the same process used in wine making where friendly yeast organisms break down the sugar from grapes to make wine. In a similar fermentation process, bacteria from the air help break down the thin coating of protein that coats each globule of coconut oil in the milky solution. The oil which separates upwards is skimmed off and gently heated. With this low heat, any water left in the oil evaporates slowly until only the virtually water-free virgin coconut oil remains in the vessel.

On top of the oil, a granular, reddish-brown meat and oil residue forms. In the Philippines, this is a local delicacy and favorite sweet called *latik.* It tastes like slightly burnt caramel, with much of the flavorful essence of the coconut. Village children are often given latik to get them out of the kitchen. Sometimes, even now as an adult, I like to have some latik pickings (imagine toasted dark chocolate, but without the sugar and oil).

One day you may become ambitious enough to want to make your own fresh coconut milk, oil, and latik. And you can do it, too! All you need are a mature coconut (the one with a brown husk), a large knife (to crack the hard nut), a grater (sharp enough to cut into the firm meat), some clean muslin cloth, a pair of strong hands (for squeezing the wet meat), and a cooking vessel (preferably a flat-bottomed, heavy skillet). Then use the fresh oil in your salad, the fresh milk in any recipe ... such as the one in Chapter 12 that calls for it—fresh latik on top of vanilla ice cream. Yum!

Dry-press + Low Heat = *Regular Virgin Coconut Oil*. In this process, the nut is cracked open and the coconut meat is removed from its shell, sliced, and grated. The meat is then subjected to low heat, often set at not more than 60°F to 80°F.

The low heat makes the meat expand, curl up, and become limp. It turns golden brown as the clear, pure, colorless oil seeps through the meat then oozes out. This dry-press method is done with automatic presses that move along the meat and its oil. After the initial pressing, the virgin coconut oil may be heated further, to just below boiling temperature, to completely remove water and to pasteurize it.

Wet-press or Dry-press Without Added Heat = *Premium Virgin Coconut Oil*. The processing of this oil takes longer. Because heat is not used, choose this oil if you really want the vitamin E and other heat-sensitive natural chemicals and antioxidants in it.

Virgin Coconut Oil Extracted from Coconut Meat by a No-heat Enzymatic Process = *Enzymatic Premium Virgin Coconut Oil* (EVCO) (294). For my studies, I used virgin coconut oil extracted from coconut meat *without* heat, using just mechanical means and a food-derived enzyme blend. The enzyme is a natural product created in a rice bran base. It has the food classification "generally regarded as safe," a grade that indicates a food acknowledged to have been used safely for many years

Laboratory analysis of the physical and chemical properties of the regular, the premium, and the enzyme-extracted virgin coconut oil were about the same. The analysis also showed that the enzymes, which dissolve only in water, were not present in the oil.

Rx: Coconuts! (The Perfect Health Nut)

I named this enzyme-extracted and totally heat-free oil "enzymatic virgin" or EVCO. We use EVCO in all our dermatological studies because I believe that there may still be unidentifiable, ultra-heat-sensitive ingredients in coconut oil that may be therapeutic.

Summary. Virgin coconut oil has three categories:

- Regular grade: when heat is used. This is the oil to use. The use of 80⁰C or higher destroys Vitamin E, possibly other antioxidants. These you can readily get from other nutritional sources.
- Premium grade: when no heat treatment is used. More expensive, less easy to find, Some traders call this Extra Virgin Coconut Oil.
- Enzymatic: also more expensive, less easy to find. No known advantages for consumer use; primarily of laboratory/studies interest.

Different Kinds of Coconut Oil

KINDS OF COCONUT OILS	Premium Virgin Coconut Oil	Regular Virgin Coconut Oil	Enzymatic Virgin Coconut Oil (EVCO)
MOISTURE CONTENT	0.12% or less	0.12% or less	0.12% or less
FREE FATTY ACIDS	0.5%	0.5%	0.5%
HARVEST	9-10 months	9-10 months	9-10 months
COCONUT TASTE	Strong	Milder	Strong
COCONUT SMELL	Strong	Milder	Strong
COLOR	Colorless	Colorless	Colorless
CLARITY	Water-clear	Water-clear	Water-clear
USE	Any kind of cooking	Any kind of cooking	Any kind of cooking; dermatological and clinical studies
PRICE	More expensive	Less expensive	More expensive
REFRIGERATOR EFFECTS	Solid in refrigerator. Liquid at room temperatures above 25°C / 76°F	Solid in refrigerator. Liquid at room temperatures above 25°C / 76°F	Solid in refrigerator. Liquid at room temperatures above 25°C / 76°F
STORAGE	Room temperature tropical or temperate	Room temperature tropical or temperate	Room temperature tropical or temperate

To reiterate, whether it is premium, regular, or enzymatic, the one difference between these virgin coconut oils is that one was processed with heat and the other was not. Vitamin E and other antioxidants might be lost during the heat processing, so just take these in your other supplements. The

EVCO has been my favorite coconut oil for testing. It is, unfortunately, not yet available in commercial quantities, but it can be customized for anyone interested in this totally heat-free, food enzyme-derived virgin coconut oil. See the Product Guide at the end of this book for assistance.

Which Grade of Coconut Oil to Use ... Superb Simplicity

Any virgin coconut oil has all the health-promoting and anticancer properties described in Part 1 of this book. Use it for cooking and for salads. The more expensive premium and the regular virgin coconut oils are the same, except for the possible loss of some antioxidants. However, you can source these antioxidants from fruits, vegetables, and other supplements. The bottom line is superbly simple: for your daily diet, any brand of regular virgin coconut oil will do.

Coconut Oil Selection is a Matter of Preference ... Coconut Oil Tasting.

When most of us began to taste and test wines, we probably took a while to develop our own personal preferences. I have my own favorite wines now, for certain meals or moods. As with your choice of wines, you will find that virgin coconut oil brands have mild but interesting differences in texture, aroma, and taste.

Develop your own preferences by trying several kinds of virgin coconut oil. Learn their different nuances, experiment with them. As you use virgin coconut oil in your meals, you'll develop your own repertoire of virgin coconut oil brands to suit your needs—in different kinds of foods and recipes, to take in a tablespoon or two.

Let's start with the aroma and taste of the oils. This often varies according to how the virgin coconut oil was made. The oils with the strongest characteristic smell and taste, and with a denser texture, are usually produced directly by dry-pressing from the coconut meat. One could describe this as *full* and *robust*. It is my favorite coconut oil. This choice, perhaps, goes along with my preference for the full-bodied taste and smell of wine, red or white.

The virgin coconut oil produced by the traditional wet-press method has a texture that is not so dense. The coconut taste and the aroma are distinct but moderate. Perhaps people will find this oil more likable, just as it seems that some people tend to prefer lighter red or white wines.

Some brands feature a virgin coconut oil that is thinner than the others and has a lighter coconut taste and smell. This is the oil made from *sapal,* the Filipino name for the protein-rich, fibrous portion of the meat left after the milk has been removed. The oil from *sapal* has the least coconut odor. Those not used to the fragrance of the coconut may find this residue-extracted, "unscented" oil desirable in their food or as an oil on their skin. A deodorized variant of enzymatic virgin coconut oil (EVCO) that I use for some patients is filtered through carbon. The texture of the filtered EVCO is denser than the very light *sapal*-derived oil.

Again, I encourage you to try many brands of virgin coconut oils, just as you probably did before you settled on your favorite wine. Whatever you eventually choose—robust, moderate, or mild; dense or light—is just a matter of personal preference. All virgin coconut oils made with care, by any method, and with virtually no water residue, are stable and will last for months, even years. And all have the health attributes you have already read about.

Copra Oil. Some people have heard about copra oil. It is not a virgin coconut oil and people tend to write unkindly about it. But it, too, has its special, even historical uses. In the early 1900s, the glycerol of coconut oil was used to replace the more expensive dairy milk in butter. Today, it is a common ingredient in soaps.

Copra is extracted from coconut meat that has been kept outdoors to dry under the sun or that has been dried in ovens. The copra is transported to processing plants, often far from the coconut's place of origin, is ground to a pulp, and then is pressed to squeeze out the oil. The basic oil processed from copra, with a press or a solvent, is crude copra oil. This is used for many industrial purposes.

The residue or copra cake is then treated with solvents to remove the remaining oil. Further refinement of this copra oil makes it suitable for use in cosmetics or in food. Alkalis are chemicals that are used to remove the free fatty acids that can make the oil rancid. Steam, under a vacuum, removes odors and flavors, while carbon filtration removes yellow or dark colors. The end result is a refined, bleached, deodorized oil, and it does *not* have the smell of coconuts. The alkali chemicals used in the processing of these oils are food grade. Like the enzymes mentioned earlier, even if these alkalis remain in the copra oil, they are classified as "generally regarded as safe." Since RBD oil is not hydrogenated, we often use RBD coconut oil for deep-frying at home.

From the refined, bleached, and deodorized oil, fractions are separated off to produce glycerin and coconut fatty acids. These, in turn, are separated by chemical means into different fractions that you can find in some of your cosmetics and skin care products.

Partially Hydrogenated Coconut Oil. Like the PUFAs, coconut oil can be subjected to partial hydrogenation by using high pressure, high temperature, and a nickel catalyst. This produces Coconut 96, which has a higher melting point of 96°F. Compared to the virgin and the other copra oils, this oil is more resistant to oxidation and has a more permanently solid texture. This is a more commercial oil, used in candies and encapsulated products.

While partial hydrogenation in the PUFAs normally produces a large amount of trans fats, it does not in this case because coconut oil has only 8% unsaturated fatty acids. These are about 6% oleic acid MUFAs and about 2% PUFAs. Roughly less than half of these, perhaps 3% to 4%, may be converted to trans fats by partial hydrogenation. This is a very small amount, compared to the 40% or more bad-for-the-heart trans fats in the partially hydrogenated PUFAs.

Organic Virgin Coconut Oil. Some people prefer to only eat organic food products, those produced without using solvents, fertilizers, or pesticides derived from nonliving or inorganic chemicals. Organic-certifying organizations impose many criteria before any food ingredient or product is allowed to make this claim. At times farmers plant cash crops—(those they can harvest and sell in just a few months after planting like vegetables even flowers or corn)—in between the coconut trees. Indiscriminate use of fertilizers and insecticides that are inorganic, pollutes and can remain for a long time in the soil to also render it inorganic

For the purist, properly certified organic virgin coconut oils, and the other products from these coconuts, are available in the organic sections of grocery stores.

The ultimate Low-carb, High-fiber Flour... Coconut Flour

In addition to coconut oil, this Perfect Health Nut produces many other foods that are healthy for our daily diets, including its meat, milk, and flour.

Coconut Meat for Low-carb Lovers. You are probably most familiar with the coconut meat in its dry, grated form (as it appears in macaroons, for example). This meat is low in calories and contains three times as much fiber, and as much protein, as vegetables. Whether eaten raw or dried, whole, shredded, grated, or dessicated, the meat is therefore an ideal ingredient or substitute in the food of those who are "low-carb"— conscious. See the comparative nutritional information in Chapter 12.

Coconut Flour = Preservative Free, Chemical Free, Low Fat, More Fibrous Than Other Flours, Diabetes Friendly, and Cholesterol Lowering. Philippine coconut research has produced another exciting new food product to use in our foods: coconut flour (295). This comes from the lowly *sapal*, which used to be thrown out for use as gruel for farm animals. The finished flour has no preservatives and is produced without chemicals. It has a cream color similar to other commercially available flours, but it is different because of its amazing and distinctive attributes. Since all the oil is removed from *sapal*, the flour produced has a low fat content. This quality and its bland favor make the flour suitable in the preparation of various food products, plus it is an excellent source of dietary fiber. For every 100 grams of this flour, as much as 60% is dietary fiber, 95% of which is soluble! This is as much as four times the soluble fiber found in other flours, including wheat, oat, or flaxseed.

The high fiber content makes coconut flour, and its flakes, ideal for health conditions reported to be improved by high-fiber foods. For example, in the control of diabetes, where the glycemic index of food is an important consideration. This index is a measure of the sugar content of food products that need insulin to break it down. The standard is sugar, which has a glycemic index of 100. Less refined carbohydrates need less insulin, thus have lower glycemic indices. Even better to the glycemic indices of food is the addition of coconut flour. Studies have shown that the glycemic indices decrease proportionate to increasing levels of dietary fiber from the coconut flour (296, 297).

Colon and breast cancer have been reported to be reduced by high-fiber food but more recent data do not appear to confirm this. Even so, the other advantages for high fiber foot cited above are very important to your health. In addition to the many desirable effects of a high fiber food already mentioned is the effect of coconut flour on cholesterol. Among patients with moderately raised blood cholesterol levels, the products made from or with coconut flour have been shown to produce a cholesterol-lowering effect (298). For those with problems of bowel regularity, high fiber foods serve as an excellent, naturally cleansing laxative.

Coconut flour has the advantage of mixing easily with both oil and water in the making of processed foods. It absorbs more water than other flours, yet is also able to mix with oils. Therefore, in the making of sausage, meatloaf, hotdogs, and other meat products, coconut flour can be used as filler, extender, and bulking agent. Similarly, coconut flour can be easily added in the making of junk food. Its high-fiber presence would give these nutritionally-scorned products some measure of healthy respectability. In all these preparations, coconut flour absorbs fat well, improving flavor retention and mouth feel. That means these products can be both more healthy and delicious!

Check out the Product Guide at the end of this book for sources of coconut flour, and start baking with it yourself with the recipes in the final chapter.

Rx: Coconuts! (THE PERFECT HEALTH NUT)

In a Nutshell

Regular virgin coconut oil

- is an excellent natural food oil for your diet;
- has an inherently very low free fatty acid content;
- is stable even when exposed to heat and is ideal for cooking;
- is produced by simple, natural means;
- produces an oil that may last for years when it is extracted through cold-pressing followed by *low heat,* to assure a low moisture content;

Premium virgin oil and EVCO

- are extracted without heat or by using a water soluble *food enzyme* (this is called enzymatic virgin coconut oil, or EVCO)
- are expensive and not readily available;
- are useful for clinical studies

All kinds of virgin coconut oil remain an all-around health food that may provide protection from cancer, is good for the heart, is trans fat free, and has multiple other benefits; and

- coming from the coconut *meat or milk* are subtly varied in aroma, taste, and texture.
- choosing the right virgin coconut oil for you is superbly simple: select according to your personal tastes.

Coconut and olive oils are both virgins but are produced by different means, judged by different standards. Both are rated excellent for health and both are distinctively delicious. If only RBD coconut oil is available where you live, use it! It still has some of the positive attributes of virgin coconut oil, namely, it is not hydrogenated and is trans fat free.

In the next chapters, you'll put it all together—the coconut's dietary, skin and overall health benefits—in an easy-to-follow "prescription" for living, the Rx: Coconuts Lifestyle. You'll have some comparative nutritional information to help you in your food selections, and you'll get treated to a bevy of healthy, tasty, coconut-based recipes. You'll soon be on your way to taking full advantage of this Perfect Health Nut.

Chapter Eleven

Rx Number 3:
The Rx: Coconuts Lifestyle . . .
A Prescription for Head-to-toe,
Cell-to-surface Health and Beauty

In This Prescription

- ☑ What's in a Lifestyle? Diet, Health, and Skin
- ☑ Fundamentals of the Rx: Coconuts Lifestyle—Dietary Recommendations: Taking the Best of the Best and Substituting Coconuts
- ☑ Fundamentals of the Rx: Coconuts Lifestyle—Health and Skin Care Recommendations: Inside-out, Outside-in Care
- ☑ The Rx: Coconuts Lifestyle
- ☑ In a Nutshell

Now you know a great deal about how—just by substituting coconut oil for other oils in your diet—you can lose weight, be heart-healthier, improve your cell membranes' structure, better your overall health, and even help protect your cells from cancer. You've learned how healthy other coconut food products are. You've also learned how coconut oil and other coconut derivatives can moisturize, treat acne, rejuvenate skin, and serve as an excellent antiseptic. It's now time to use all this knowledge to enhance how you *live:* your diet, health, and skin . . . inside-out, head-to-toe, cell-to-surface health and beauty. The good news is that it's easy. In this prescription, the Rx: Coconuts Lifestyle, you get a simple how-to guide for integrating coconuts into your life.

What's in a Lifestyle? Diet, Health, and Skin

What's in a lifestyle? Why a *lifestyle*? Why not just a diet or regimen? I actually started out by formulating a comprehensive diet and other regimens using the coconut. As I continued to research and write, however, I realized that the benefits of the coconut apply to so many aspects of our overall health that no one diet or regimen could take advantage of them all. I soon saw that it wasn't only a step-by-step regimen that was needed but also a whole attitude change toward the coconut—its substitution for many of our foods and its use in many other aspects of our lives. It was "Think coconuts!" "Have coconuts on the brain," "Choose to live a coconut-focused *lifestyle*."

Some may find this sacrilegious, looney, or both, to the point of thinking the good doctor has come down with a tropical fever and gone coco-loco. Seriously though, folks, through these pages I have tried to present a sane, science-based exposition of the coconut together with the other fats and oils in our diet. Obviously I'm not suggesting that the coconut become a panacea. Neither am I talking about a cult, a fad or even a spiritual transformation.

A coconut-focused lifestyle is for the thoughtful person who, having learned the science, albeit simplified, can now have a handy and positive point of reference by which to achieve that larger balance we seek in our daily lives. By thinking coconuts based on its science, we can then make small yet consistent changes in our lives that can have a longer impact on our overall well being

So what is the main difference between a lifestyle and, say, a diet or a regimen? They're very similar, obviously. Both entail making choices to help yourself achieve a specific goal or set of goals. The main distinction is that whereas a diet or regimen usually entails following specific steps, a lifestyle is a less structured—but a consistent and committed—approach toward achieving the same goals. A lifestyle is a way of being, a mind-set, a way of thinking. Instead of following set steps, living a particular lifestyle means its basic tenets, principles, and goals inform all or most of your choices, in all or most areas of your life—from food, to skin, to overall health, and in some cases, to clothing, cocktails, and clubs!

I think of it this way: I can go on an organic diet. This could mean that I will eat only organic food and drink only organic beverages. However, if I choose to *live an organic lifestyle,* not only would my food and drinks be organic but also so would my clothes, and maybe my furniture and my pet food. I'd possibly ask around for organic health supplements and look for certifiably organic medicines, skin care products, shampoos, deodorants, and even house-cleaning, gardening, or laundry products.

As a scientist and doctor, ever curious but skeptical, I'd find my eye being drawn to articles about organic goods and news on organics in general. I'd be constantly open to learning, yet questioning what I'm hearing so as to guard against unscientific fads and frauds. Everything I did, whatever I bought, the countries I chose to travel to, the restaurants or bars I'd go to . . . almost everything in life and most of my choices would have an organic "vibe" or fit in with my organic philosophy. I might even be drawn to new friends who were organic-minded. While I would try to avoid becoming a tiresome fanatic to my old friends, they would certainly get an earful about my new organic lifestyle.

So why do I recommend a coconut-focused lifestyle instead of a coconut-focused diet or regimen? Four reasons: inside-outside-inside health; all-around commitment to health; simplicity; and a bonanza of health benefits.

First: Inside-Outside-Inside Health. In many ways, the human body's inside and its outside reflect each other's current state. For instance, when we're fighting the flu, we tend to look gaunt and noticeably pale—signaling to the outside world, "I'm weak, please be careful with me . . . and bring me some chicken soup, quick!" Similarly, an external "ugliness" can point to an internal problem. Boils or sores on the skin (external infections themselves), for example, can indicate an internal loss of immunity or a low-grade infection like strep throat.

I firmly believe that the healthier you are inside, the better you look outside. As a dermatologist and a practicing physician for many decades, I have seen this proven countless times. The reverse is true, too: looking good *naturally* (without the endless effort of hiding pimples or wrinkles with makeup or of concealing bulges with "camouflage dressing") can give you a sense of confidence that, I am convinced, helps you feel calmer, more centered, happier to be you. All of this can help lessen stress, which, as scientific studies are increasingly showing, plays a vital role in causing sickness and the body's degeneration.

Perhaps the idea of lessening stress by naturally looking good sounds vain or superficial. But I'd wager that a great many of us have, at one point or another, experienced the anxiety and insecurity—stress—that come with a poor self-image. On the flip side, when we're not stressed about the way we look, we enjoy everything more—shopping for clothes, going out to parties, even sex. "Comfortable in your own skin" is the best way to put it, with *comfortable,* i.e., stress-free, being the operative word.

Unlike a diet or a skin care regimen, therefore, the Rx: Coconuts Lifestyle focuses on the health of both your inside *and* your outside. It addresses (for your inside) the health side of weight control, your heart, your cells' membranes, and the fight against free radicals and internal infections; and (for your outside) the aesthetic side of weight control, acne therapy, moisturization, rejuvenation, and the battle against skin infections.

Second: All-around Commitment to Health. My second reason for prescribing a coconut-focused lifestyle instead of a diet or regimen is that being healthy is a commitment to improving *all* aspects of your life, not just

Rx: Coconuts! (The Perfect Health Nut)

what you eat. The Rx: Coconuts Lifestyle encourages you to substitute coconut oil and food products as often as you find comfortable and easy for your regular oil, flour, and butter, for example, but also to use coconut derivatives in your daily health and personal care routines. This means committing to healthier meals, nutritional supplements, antiseptics, basic hair and body care, skin treatments, even what you use as a health drink after sports!

Third: Simplicity. Successfully adhering to a diet, skin care, or personal care regimen is not easy. There's a lot to remember: percentages and grams and doses and phases. In addition, because *regimen* or *diet* implies rules, rules, rules, the times you stray (tellingly termed *cheating*) can be demoralizing and can cause you to give up in frustration. And consistency and balance are key ... you do not want to jump from one fad diet or regimen to another. Once you've found the right one, persisting with it in reasonable and balanced fashion gives you the best results. If you've managed to do that, you should give yourself a real pat on the back ... and, importantly, not make it harder for yourself to keep on sticking to it!

Don't misunderstand; health entails discipline. You *must* see your doctor for sicknesses and infections, for proper skin consultations, to monitor your heart and your overall health. You *must* get regular checkups and blood tests. You *must* exercise and eat according to the diet prescribed to you by your physician. This is, after all, the same disciplined care you extend to your pets or even to your car!

My point here is that, instead of creating yet another diet or fanatical regimen for you to learn and follow, I designed the Rx: Coconuts Lifestyle to *work with* the diet and regimen that you already have, that already works for you, that you're already sticking to. By making coconuts a *lifestyle* choice, you simply *substitute* coconuts as much as you reasonably can into your existing diet (still following your diet's rules and calorie requirements) and *add* coconuts and its derivatives to your personal care regimen and daily routines. Simple.

Fourth: A Bonanza of Health Benefits! The coconut and its derivatives are so good for so many aspects of your overall health that it would be a shame not to enjoy as many of them as you can! Living the Rx: Coconuts Lifestyle—more than following one coconut-based diet or regimen—is the best way to take advantage of as many of these healthful benefits as possible.

As you'll see further on, the Rx: Coconuts Lifestyle is a "prescription" or series of suggestions for food and health supplements as well as for skin and personal care. To get a better understanding of what the Rx: Coconuts Lifestyle helps you achieve, let's explore the fundamentals of its recommendations.

Fundamentals of the Rx-Coconuts Lifestyle—Dietary Recommendations: Taking the Best of the Best and Substituting Coconuts

The dietary recommendations in the Rx: Coconuts Lifestyle are simple: (a) for your cooking, baking, and food preparation, selectively *substitute* coconut oil and coconut flour for other oils or flours; (b) instead of using regular, high-calorie butter, selectively *substitute* coconut milk or cream; and c) *selectively add* other coconut foodstuffs and coconut water. Do this while staying with the healthy and effective diet you prefer and while adhering to the recommended caloric intake set by your physician (or as specified by your diet) for your particular body type, weight, and weight goals.

These dietary suggestions are based on all the health benefits discussed earlier in this book; on current recommendations in nutrition, such as those of the U.S. National Research Health Council (299) and Harvard's School of Public Health (300); and on some of the best new things we have learned from the successful and effective diets of Drs. Agatston and Atkins.

New Dietary Recommendations Recognize Fat's Importance. Let's look at the newly published recommendations from the United States National Research Health Council's Committee on Diet and Health for 2005. It specifies that between 20% and 35% of our food should be composed of fat; now because we understand its meaning, we can say: "good fat." This is a tasty diet! This recommendation is *up* from the previous maximum of 30%.

In the past, there was no recommended minimum for fats. The Council currently specifies a minimum of 20%. This change clearly recognizes the fact that fat in the diet *is* important. It is unhealthy to go ultra-low fat. Remember that with ultra-low-fat diets, you risk, among other things, too many PUFAs in your system, which have hidden trans fats and hidden carbs, and which can affect your cells' membranes and their susceptibility to free radicals. If you happen to be a strict low-fat eater, take note of this.

Now, which fats to eat?

Sadly, yet not surprisingly, coconuts still are not included in these recommendations (expect this to change, however, as the new research on coconut oil makes it to the mainstream). Coconut oil *is* saturated, but you now know that it is different from the cholesterol-rich long-chain saturated fat from animals and dairy. Among other things, coconut oil is medium chain and cholesterol free. It's also trans fat free. You have also learned from previous chapters that coconut oil has antiobesity, heart-healthy, and possible anticancer attributes.

Let's put together what you now know about virgin coconut oil and the recommended amount of fat that's right for you in the dietary part of the Rx: Coconuts Lifestyle. It's simple. Make sure that your intake of oil includes healthy virgin coconut oil.

Atkins and South Beach. Both these diets drew our attention to protein, fat, and fiber and away from carbohydrates. The South Beach diet emphasizes eating healthy fats from lots of seafood, olive oil, nuts, and avocados. However, it says no to *all* saturated fats, does not differentiate between them, and does not talk about coconut oil. Dr Atkins's diet also gives no set fat allowances and in fact encourages the eating of more fat, of all kinds, at least at the start. As you get going, both diets caution moderation in the amounts, or as you become full.

While both diets are still being studied by nutrition experts, there are sufficient successes, perhaps like you, for the low-carb diets and foodstuffs that now are bywords among dieters and in the marketplace. As

I mentioned earlier, I, too, consider myself a success story of both diets . . . but now, with coconut oil!

The dietary recommendations of the Rx: Coconuts Lifestyle work well with your current weight-loss and/or heart-healthy diet, including both the Atkins and South Beach (in all phases of the diets). Again, in the Rx: Coconuts Lifestyle, you stay with your diet, and as much as you find you enjoy, selectively substitute coconut oil for your oil and fat needs, coconut flour for your flour needs, coconut milk or cream for your butter needs, and selectively include other coconut foods and coconut water. This makes your existing diet expand its goals from just weight loss and heart health to cancer prevention and (potentially) to skin enhancement and disease prevention.

How exactly do these dietary recommendations of the Rx: Coconuts Lifestyle help achieve all these goals?

I suggest that if you are trying to lose weight, and your cholesterol values are above normal, that you go closer towards the lower (20%) recommended amount for fat in your diet. Go higher, toward the 35% fat in your diet if your weight is steady from balancing this amount of fat with a physically active life.

Dietary Goals of the Rx: Coconuts Lifestyle . . . *Weight Loss.* You achieve weight loss with the Rx: Coconuts Lifestyle by the simple substitution of virgin coconut oil in your cooking and food needs. Remember from Chapter 1 that, compared to other food oils, coconut oil—predominantly made up of medium-chain fatty acids—is the best oil to help you stay lean because it has fewer calories, avoids the general circulation (getting to the liver more quickly), burns energy faster than other oils, and satisfies your hunger while you eat less. Check out the comparative information in Chapters 1 and 12. The numbers speak for themselves.

Do note that, just as in any diet, weight loss or not, you *should still watch your total calorie intake.* A common recommendation is 2,000 calories per day, less to lose weight. This recommendation is different for each person, depending on weight, frame, and weight-loss goals. As you substitute coconut oil into your diet, read your other foods' labels and learn their nutritional information, to keep track of the calories you're ingesting. Coconut oil has somewhat fewer calories than other oils, but you need to balance your overall caloric intake from your other foods, too, especially if you're trying to lose weight. Calorie-count.com is a great source for learning about the calories in your foods—I found it to be one of the more complete and easy to use.

Dietary Goals of the Rx: Coconuts Lifestyle . . . *Heart Health*. Coconut oil helps you achieve heart health because coconut oil is cholesterol free, is low in carbohydrates, is free of hidden carbohydrates, is low in linoleic acid—to control the risk of blood clotting—and, vitally, is trans fat free. We've seen coconut oil's efficacy for heart health in historical data. In Chapter 3, we saw how populations from coconut-eating countries are heart-healthier than those from non-coconut-eating countries, and that when we eat coconut oil, our good HDL cholesterol goes up while the bad LDL remains the same.

Dietary Goals of the Rx: Coconuts Lifestyle . . . *Cancer Protection*. We learned in Chapter 4 that, in addition to the other important oil ratios you should keep in mind, diets high in coconut oil seem to have a positive effect against cancer. While more research is needed, it is well-established that coconut oil does keep your cell membranes healthier and more stable, and keeps their structure balanced. This helps keep them more resistant to invasion by the free radicals that can be a prelude to cancer.

Chapter 4 also showed that whereas olive oil has been shown to prevent the promotion and spread of cells that are *already* transformed into cancer cells, initial studies show that coconut oil can act even *earlier*, at the *initiating phase* of cancer growths (in that study, coconut oil proved to be antigenotoxic), which indicates that it may help prevent the *formation* of cancer cells.

To help you achieve these dietary goals, just follow the Rx: Coconuts Lifestyle's ultra-simple dietary recommendations: wherever feasible, selectively substitute virgin coconut oil for your oil and fat needs, coconut flour for your flour needs, coconut milk or cream for your butter needs, and selectively include other coconut foods and coconut water.

Fundamentals of the Rx: Coconuts Lifestyle—Health and Skin Care Recommendations: Inside-out, Outside-in Care

The Rx: Coconuts Lifestyle goes beyond dietary recommendations. As you will see, by keeping "coconuts on the brain," you apply the advantages of this Perfect Health Nut to your skin and overall health.

Inside-outside Health and Hygiene. In addition to including coconut oil and food in your diet, in Chapter 9 we learned of an easy way to deliver all the coconut's rewards directly into your system: drink a tablespoon of it every day. As you take your daily bundle of vitamins and antioxidants, make the last swallow a tablespoon or two of virgin coconut oil.

If you keep sore soothers or throat lozenges in your medicine cabinet, keep some coconut oil handy, too. You can gargle with it when your throat starts getting itchy or when you feel mouth sores coming on.

The coconut derivative monolaurin is an antiseptic and disinfectant that is clinically proven to be as effective as alcohol, without being irritating to the skin. It's been shown to be effective in preventing infections as well as treating already-infected areas of the skin. Having a bottle of monolaurin handy can help keep at bay the infectious bugs that you can pick up throughout your daily activities. A bottle in your medicine cabinet is great for common nicks and scratches or for when muddy kids come in from soccer practice a little cut up. Monolaurin is also ideal if you travel quite a bit or work in high-risk environments such as hospitals or clinics where constant disinfection is recommended.

Skin. In Chapter 6, we saw how monolaurin also helps treat acne and sweat acne, and how it can lessen shine. This is a nonirritating, simple, and easy option for those who suffer from very oily skin or pimples. Just add monolaurin to your regular anti-acne therapy to enhance its efficacy and to control shine . . . without adding dryness, sensitivity, or irritation. It makes oily skin feel clean and refreshed, too!

If your skin is the opposite—very dry—the encouraging clinical results regarding virgin coconut oil show that its fatty acids are native to the skin's barrier and are therefore absorbed better and moisturize more thoroughly. New studies further show the potential of coconut oil to deliver its antiseptic and antioxidant properties to the skin as it moisturizes. Skin feels and looks impressively soft and smooth while you are enjoying coconut oil's myriad health benefits.

As you buy your personal care and skin care products, start looking for monolaurin and other coconut derivatives. Monolaurin as a preservative is a nonallergenic alternative to traditional cosmetic preservatives—excellent news for the very sensitive. Plus, it has the potential to enhance the product with its antiseptic attributes. As more research is undertaken, make sure to look out for growth hormones from the coconut. Coconut-derived kinetin has already shown promise in the field of skin rejuvenation, and studies are under way to search for more growth hormones and more applications for our skin and hair.

Now that we've covered the fundamentals, let's get to it! The Rx: Coconuts Lifestyle appears below. Don't forget to check out the nutritional information and comparative values in Chapter 12, as well as the Product Guide at the end of this book. These—and the tasty, coconut-based recipes—will help jump-start your Rx: Coconuts Lifestyle.

The Rx: Coconuts Lifestyle

A Prescription for Head-to-toe, Cell-to-surface Health and Beauty

Helpful data appears in the Nutritional Information and Comparative Values sections of Chapter 12. For a list of different places around the world where you can find many of the foods and products mentioned in the following pages, see the Product Guide at the end of this book.

The Rx:Coconuts! Lifestyle
A Prescription for Head-to-toe, Cell-to-surface Health and Beauty

Diet (Cooking, Baking, Food Preparation)

What: **Virgin Coconut Oil**

Why: Cholesterol Free, Cancer Protection, Heart Health, Cell Membrane Health, Trans Fat Free, Weight-Loss.

How: Substitute it for any oil needed for cooking, as a salad dressing or mixed with salad dressing, as a butter or dip, or in baking.

Substitute for other oils ☑   **Per calorie needs** ☑
Per your current diet's rules ☑   **When needed** ☑

Remember: for weight loss, always count your total calories.

Other: Virgin coconut oil works surprisingly well in most recipes. If you prefer less coconut flavor, try deodorized virgin coconut oil or infuse it with herbs and spices (Chapter 12). For very particular recipes, use another healthy oil and sprinkle in some coconut oil (or use it in another part of the meal). The goal is to substitute coconut oil for your regular oil any time you need an oil for cooking or food preparation.

Diet (Baking, Flour Needs)

What: **Coconut Flour**

Why: Cholesterol Lowering, Chemical Free, Diabetes Friendly, Low Fat, More Fiber than other flours, Preservative Free

How: For baking. For making meat products. Use as a substitute whenever flour is needed.

Substitute for other flours ☑   **Per calorie needs** ☑
Per your current diet's rules ☑   **When needed** ☑

Check out chapter 12 for tasty coconut-based recipes.

Other: Coconut flour mixes well with both oil and water. Use it for baking, or as a filler, extender, and bulking agent when making sausages, meatloaf, hotdogs, and other meat products ... basically, use coconut flour as a substitute whenever flour is called for.

there's more...

Rx: Coconuts! (The Perfect Health Nut)

The Rx: **Coconuts!** Lifestyle
A Prescription for Head-to-toe, Cell-to-surface Health and Beauty

Diet [Health Drink]

What: **Coconut Water**

Why: Dye Free, High Electrolyte, High Fiber, High Magnesium, High Potassium, High Sodium, Low Acidity, Low Calorie, Low Carb, Low Sugar, Natural, Preservative Free, Unprocessed.

How: Take as a healthy, tasty, refreshing sports drink.

Substitute for other oils ☑ **Per calorie needs** ☑
Per your current diet's rules ☑ **When needed** ☑

Coconut water versus other sport drinks? Comparative Values in chapter 12.

Other: Best when fresh. Great to drink daily as a refreshing juice, or after sports or working out. In some countries, more convenient snack-drink packs are available. For these snack drinks, check that they don't have sugar or other additives.

Diet [Cooking, Snack]

What: **Coconut Meat**

Why: All the benefits of coconut oil and water.

How: As an ingredient in cooking, to sprinkle over deserts, or fresh as a healthy snack.

In cooking, as nut substitute ☑ **Per calorie needs** ☑
Per your current diet's rules ☑ **When needed** ☑

Check out chapter 12 for tips on how to select, store and use coconut meat, and for carb content compared to other nuts.

Other: Available fresh, unsweetened, shredded, dried (desiccated), flaked, toasted, creamed. For low-carb diets, make sure you use unsweetened coconut meat. Coconut meat has all the benefits of coconut oil and water. It is ideal for special snacks and recipes in lieu of other nuts that have much higher carb-counts, but not necessarily for daily consumption.

there's more...

The Rx:**Coconuts!** Lifestyle
A Prescription for Head-to-toe, Cell-to-surface Health and Beauty

Diet [Asian Cooking, Butter Substitute]

What: **Coconut Milk or Coconut Cream**

Why: Cholesterol Free, Less Calories Than Butter, Flavor

How: Use as needed (for some Asian recipes like curries), as a butter substitute or to lessen overall butter content (such as in mashed potatoes).

Substitute for butter ☑ Per calorie needs ☑
Per your current diet's rules ☑ When needed ☑

Other: Rich and flavorful. For use in popular Asian recipes. Coconut milk gives you many benefits of the coconut and its calorie count is much lower than butter. Substitute for (or mix with) butter for a lower calorie count in your meals. See the table in Chapter 1 for comparative values.

Nutritional Supplement [Daily Health]

What: **Virgin Coconut Oil**

Why: Helps fight infections, helps skin look good from the inside-out, a mild laxative, delivers all the benefits of coconut oil directly to your system.

How: Drink 1-2 tablespoons daily.

1x-a-day ☑   Twice-a-day ☐   As often as possible ☐
AM ☐ PM ☐   When needed ☐   As often as needed ☐

Coconut oil naturally switches from liquid to butter. Chapter 12 has storage and use tips.

Other: Most people can drink virgin coconut oil directly (if your coconut oil has "buttered" simply re-liquify and drink). If you're one of the few who has difficulty drinking the oil directly, try taking it with a spoonful of honey, drizzled on your dessert or as a salad dressing. Other suggestions appear in Chapters 9 and 12.

there's more...

Rx: Coconuts! (The Perfect Health Nut)

The Rx:**Coconuts!** Lifestyle

A Prescription for Head-to-toe, Cell-to-surface Health and Beauty

Nutritional Supplement (Daily Health)

What: **Coconut Oil Capsules**

Why: Helps fight infections, helps skin look good from the inside-out, a mild laxative, delivers all the benefits of coconut oil directly to your system.

How: 1-2 capsules, 3x-a-day.

1x-a-day ☐	**3x-a-day** ☑	As often as possible ☐
AM ☐ PM ☐	When needed ☐	As often as needed ☐

Other: Each capsule contains 1 gram of coconut oil whereas a tablespoon gives you 15 grams of coconut oil. I really recommend drinking 1-2 tablespoons of virgin coconut oil a day. However, if you are one of the few who really cannot drink coconut oil, taking coconut oil capsules, while not as good as the 15 grams of the oil, is better than nothing!

Skin (Antiseptic, Antibiotic)

What: **Monolaurin Antiseptic Gel**

Why: Very effective, non-skin-irritating antiseptic for frequent use.

How: Apply as often as necessary on hands and feet. Use as an antiseptic to disinfect wounds or several times a day to prevent infections.

1x-a-day ☐	2x-a-day ☐	As often as possible ☐
AM ☐ PM ☐	When needed ☐	**As often as Needed** ☑

Other: This is such a simple yet incredibly effective way to combat the bugs that you pick up on a daily basis. Keep the gel in your bag, briefcase, or purse, and also when you travel, for quick disinfections. Will not dry out or irritate skin.

there's more...

The Rx: Coconuts! Lifestyle
A Prescription for Head-to-toe, Cell-to-surface Health and Beauty

Skin Treatment Aid For Acne / Sweat Acne

What: **Monolaurin Gel**
Why: Helps prevent acne and control shine.
How: Use with your regular acne treatment. Or, use under make-up or on bare skin to control shine. Use on back, chest and body skin to help clear up sweat acne.

1x-a-day ☐ **2x-a-day** ☑ As often as possible ☐
AM☑ **PM**☑ When needed ☐ **As often as Needed** ☑

Other: Enhances the efficacy of your regular acne treatment. To use with your acne therapy, apply after your acne astringent or before your acne cream. To control shine, apply under make-up or on bare skin.

Skin (Cosmetic Preservative)

What: **Monolaurin**
Why: Non-allergenic preservative that delivers all the other benefits of monolaurin.
How: Try to shop for cosmetics, skin care, hair care, shaving gear, bath stuff and other personal care supplies that are preserved with monolaurin.

1x-a-day ☐   2x-a-day ☐   **As often as possible** ☑
AM ☐ PM ☐   When needed ☐   As often as needed ☐

Other: Using only truly hypoallergenic (preferably validated hypoallergenic), non-comedogenic products helps prevent a host of skin problems including rashes, irritations, dark spots, and acne. Monolaurin-preserved cosmetics reduce the risk of irritations from traditional preservatives and give you acne-fighting, disinfecting, and other benefits from this excellent coconut-oil derivative.

there's more...

The Rx:**Coconuts!** Lifestyle
A Prescription for Head-to-toe, Cell-to-surface Health and Beauty

Skin (Moisturizer)

What: **Virgin Coconut Oil**

Why: An excellent moisturizer with fatty acids that are native to the skin's barrier layer.

How: Use as your daily moisturizer or body oil. Use as an after-sun oil to help your skin recover from sun exposure.

1x-a-day ☐   **2x-a-day** ☑   As often as possible ☐
AM ☑ **PM** ☑   When needed ☐   **As often as Needed** ☑

Other: Used as a moisturizer, virgin coconut oil leaves skin soft and smooth. It also has the potential of imparting antiseptic and antioxidant properties. When used after sun exposure, virgin coconut oil may help repair damage (it also gives skin a glowing tropical sheen). Pour out a palmful (in its liquid form) or scoop out (in its coco butter form) and massage liberally over skin.

Skin (Cosmetic Active Ingredient)

What: **Coconut Kinetin Growth Factor**

Why: Initial studies show great potential for stimulating cells and rejuvenating skin.

How: If you're interested in skin rejuvenation, try to look for a treatment with coconut kinetin.

For where to buy any of these products, see the Product Guide at the end of this book.

1x-a-day ☐   2x-a-day ☐   **As instructed** ☑
AM ☐   PM ☐   When needed ☐   As often as needed ☐

Other: While more studies need to be done, the initial studies on kinetin and skin rejuvenation are promising especially when combined with micro-exfoliants and other anti-photoaging active ingredients.

Vermén M. Verallo-Rowell, M.D.

In a Nutshell

The Rx: Coconuts Lifestyle is a simple, easy-to-follow "prescription" for healthier living.

- Its dietary suggestions incorporate some of the recommendations from Harvard's School of Public Health (302), the U.S. government's nutrition groups, and the Atkins and South Beach diets, while maximizing the benefits from virgin coconut oil.
- Its dietary suggestions are designed to be easy to follow, to deliver maximum benefits with minimal relearning. Simply *selectively substitute* (a) coconut oil for regular oils and fats; (b) coconut flour for your flour needs, and coconut milk or cream for your butter needs; and (c) *selectively include* other coconut foods and coconut water.
- It increases virgin coconut oil to about one-third of the fat in your daily diet, to take full advantage of the oil's weight-loss, heart-healthy, and cancer-preventing qualities.
- It makes virgin coconut oil a daily health supplement; drinking it delivers the benefits of virgin coconut oil directly into your system.
- It plugs coconut oil and its derivatives into your skin, personal care, and hygiene routines—boosting their efficacy and adding the coconut's anti-acne, moisturizing, antiseptic, and nonirritating properties to your regimens.

What's next? Get started! You have your prescriptions, so go out and experiment with virgin coconut oils, coconut milk, and cream. Try some fresh coconut water, get some monolaurin . . . and try some tasty coconut-based recipes!

The recipes that follow are easy enough for those of you who may have bought your first bottle of virgin coconut oil to drink or to moisturize your skin, but still haven't a clue how to use it in your food. For those of

Rx: Coconuts! (The Perfect Health Nut)

you who are accomplished chefs, you'll find the selection creative and diverse, and the recipes may even inspire you to create your own coconut concoctions. In any case, the dishes are divinely delicious and can be easily incorporated into your Rx: Coconuts Lifestyle. Enjoy!

Chapter Twelve

**Rx Number 4:
Comparative Nutrition Guide and
Tasty Recipes**

In This Prescription

- ☑ Comparative Nutrition Charts: Virgin Coconut Oil and Other Dietary Oils, Coconut Water and Other Beverages, Coconut Meat and Other Nuts, Coconut Flour and Other Flours, Other Coconut Food Products Versus Butter and Mayonnaise
- ☑ Tips: Where to Look, What to Look for, Serving Suggestions
- ☑ Tips: Using Virgin Coconut Oil in Your Food on a Daily basis
- ☑ Tasty Coconut Recipes
- ☑ Beyond the Recipes: Sustained Health for Life
- ☑ In a Nutshell

In this chapter, you'll find comparative nutritional information to help guide you as you make your coconut selections. There are also tips for storing and serving your coconut products and, the tastiest part, some wonderful, flavorful recipes for you to try. At the end of it, take a quick gander at the Product Guide at the end of this book to learn where you can find the products and foods mentioned throughout the previous Chapters.

Comparative Nutrition Chart:
Virgin Coconut Oil and Other Dietary Oils

	Coconut	Olive	Canola	Soybean Salad or Cooking (Hydrogenated)	Corn Salad Or Cooking	Salmon Fish Oil
SIZE	1 cup 218 g	1 cup 216 g	1 cup 218 g	1 cup 218 g	1 cup 218 g	1 cup 218 g
CALORIES	1,879	1,909	1,927	1,927	1,927	1,966
FROM FAT	1,879	1,909	1,927	1,927	1,927	1,962
TOTAL FAT	218 g	216 g	218 g	218 g	218 g	218 g
Saturated	188.6 g	29.1 g	15.5 g	32.5 g	27.7 g	43.3 g
PUFA	3.9 g	21.6 g	64.5 g	82 g	128 g	87.9 g
MUFA	12.6 g	159.6 g	128.14 g	93.7 g	52.8 g	63.3 g
CHOLESTEROL	0 mg	0 mg	0 mg	0 mg	0 mg	1057 mg
SODIUM	0	6 mg	0	0	0	0
TOTAL CARBOHYDRATE	0	0	0	0	0	0
PROTEIN	0	0	0	0	0	0

Source: Calorie-Count.com 2003-2005

* Note the slightly smaller amount of calories in coconut oil compared to the other oils. For more details, see Chapter 2.

** Note that coconut oil's high saturated fat and low PUFA and MUFA fat means that it does not need partial hydrogenation and is therefore trans fat free. Remember, too, that coconut oil is cholesterol free and has low linoleic acid content (which can encourage the production of blood-clotting elements).

Rx: Coconuts! (THE PERFECT HEALTH NUT)

Comparative Nutrition Chart: Coconut Water and Other Beverages

	Coconut Water	Orange Juice	Grapefruit Juice	Pineapple Juice	Gatorade-Lemonade Flavor
Size	1 cup 240 g	1 cup 248 g	1 cup 247 g	1 cup 250 g	1 cup 240 g
Calories	46	112	96	140	50
From Fat	4	4	2	2	0
Total Fat	.5 g	.5 g	.2 g	.2 g	0 g
Saturated	0.4	0	0	0	0
PUFA	0	0.1	0	0.1	0
MUFA	0	0.1	0	0	0
Cholesterol	0	0	0	0	0
Sodium	252 mg	2 mg	2 mg	3 mg	110 mg
Total Carbohydrate	8.9 g	25.8 g	22.7 g	34.5 g	14 g
Dietary Fiber	2.6 g	.5 g		.5 g	0 g
Sugar	6.3 g	20.8 g		34 g	14 g
Protein	1.7 g	1.7 g	1.2 g	.8 g	0 g
Vitamin A	0%	10%	22%	0%	0%
Calcium	6%	3%	2%	4%	0%
Vitamin C	10%	207%	156%	100%	0%
Iron	4%	3%	3%	4%	0%
Manganese	.336 mg	0.025 mg	0.049 mg	2.475 mg	0 mg
Magnesium	60 mg	27 mg	30 mg	33 mg	0 mg
Potassium	600 mg	496 mg	400 mg	335 mg	34 mg
Riboflavin	.12 mg	0.074 mg	0.049 mg	0 mg	0 mg
Thiamine	0.072 mg	.223 mg	0.099 mg	.05 mg	0 mg

Source: Calorie-Count.com 2003-2005.

Note: Coconut water is ideal as a sports drink. Compared to Gatorade, it has higher sodium, potassium, and magnesium. Coconut water has about the same amount of sugar as Gatorade but with more fiber. Compared to other fruit juices, coconut water is lower in carbohydrates. UHT packs of this water is now available. Watch out for them at your local stores!

Comparative Nutrition Chart:
Coconut Flour and Other Flours

	Soy, Low Fat	Coconut	Wheat, Whole Grain	Wheat, White. All-Purpose	Rye, Medium	Corn, Enriched, Yellow
Size	1cup, stirred 88g	1 cup 100 g	1 cup 120 g	1 cup 125 g	1 cup 102 g	1 cup 114 g
Amount						
Calories	327	438	407	455	361	416
From Fat	53	117	20	11	16	39
Total Fat	5.9	13 g	2.2	1.2	1.8	4.3
Saturated	.8	0	.4	.2	.2	.6
PUFA	3.3	0	.9	.5	.8	2
MUFA	1.3	.3	.3	.1	.2	1.1
Cholesterol	0	0	0	0	0	0
Total Carbohydrate	33.4 g	68 g	87.1 g	95.4 g	79 g	86.9 g
Dietary Fiber	9.0 g	60 g	14.6 g	3.4 g	14.9 g	—
Soluble	9.0 g	3.5 g	0 g	0 g	0 g	
Insoluble	0 g	56.5 g	0 g	0 g	0 g	
Sugar	0 g	0 g	.5 g	.3 g	1.1 g	—
	17.4 g					
Protein	40.9 g	12 g	16.4 g	12.9 g	9.6 g	10.6 g
	Manganese Magnesium H Phosphorus H Potassium H		Selenium VH Manganese VH	Selenium H Thiamin H	Manganese VH Selenium H	Iron H Niacin H Thiamin VH

From: nutrition facts.com; SunGee Premium Coconut Flour

Low-fat soy flour has the lowest calories and total carbohydrate content among the flours and is therefore valuable for low-calorie, low-carb diets. Coconut flour has the next-lowest carb content. And compared to all flours, including soy, coconut flour has much, much more fiber.

Comparative Nutrition Chart:
Coconut Meat and Other Nuts

	Coconut Meat	Walnuts	Cashew Roasted in oil, no salt	Mixed Nuts	Pecans	Peanuts Dry-Roasted, no salt	Peanuts, Oil Roasted with salt
Size	1 cup raw meat 100 g	1 cup shelled (50 halves) 100 g	1 cup 129 g	1 cup 137 g	1 cup 119 g	1 cup 146 g	1 cup 144 g
Calories	354	654	748	814	822	854	862
From Fat	0	546	576	634	771	607	633
Calories Per g	3.54	6.54	5.8	5.9	6.9	5.8	6
Total Fat	33.49	65 g	62 g	70.5 g	85.6 g	73 g	76 g
SAT	0 g	6 g	11 g	9.4 g	74	10 g	12 g
PUFA	0 g	0 g	0 g	14.7 g	25.7	0 g	0 g
MUFA	0 g	0 g	0 g	43 g	48.6 g	0 g	0 g
Cholesterol	0	0	0	0	0	0	0
Sodium	16 mg	2 mg	17 mg	16 mg	0	9 mg	46.1 mg
Total Carbohydrate							
Dietary Fiber	15.23 g	14 g	39 g	34.7 g	16.5 g	31 g	22 g
Sugar	9 g	7 g	4 g	12.3 g	11.4 g	12 g	14 g
	6.23 g	3 g	6 g	—	4.7 g	6 g	6 g
Protein	3.33 g	15 g	22 g	23.5	10.9 g	35 g	40 g

Rx: Coconuts! (THE PERFECT HEALTH NUT)

Comparative Nutrition Chart: Other Coconut Food Products Versus Butter and Mayonnaise

	Coconut Meat Fresh, Unsweetened (Shredded Dried or Desiccated, Flaked, Toasted, Creamed)	Coconut Milk (Canned, Frozen, UHT)	Coconut Cream (Canned, Frozen, UHT)	Butter	Mayonnaise-Type Salad Dressing, regular with salt
Size	1cup shredded	1 cup	1 cup	1 cup	1 cup
	80 g	240 g	240 g	227 g	235 mg
Calories	283	552	792	1628	917
From Fat	241	515	749	1628	706
Total Fat	26.8 g	57.2 g	83.2 g	184.1 g	78.5 g
Saturated	23.8 g	50.7 g	73.8 g	116.6 g	11.5 g
PUFA	.3 g	.6 g	.9 g	6.9 g	42.3 g
MUFA	1.1 g	2.4 g	3.5 g	47.7 g	21.1 g
Cholesterol	0 mg	0 mg	0 mg	488 mg	61 mg
Sodium	16 mg	36 mg	10 mg	25 mg	1671 mg
Total Carbohydrate					
Dietary Fiber	12.2 g	13.3 g	16 g	.1 g	56.2 g
Sugar	7.2 g	5.3 g	5.3 g	0 g	0 g
	5.0 g	8 g	0 g	.1 g	15.0 g
Protein	2.7 g	5.5 g	8.7 g	1.9 g	2.1 g

Tips: Where to Look, What to Look For, Serving Suggestions

If you live in a non-tropical country, you can find the coconut products needed for these recipes at organic, international, Asian, or other specialty food sections. Or order online through the web stores specified in the Product Guide at the end of this book.

The coconut products you can easily find (even in regular supermarkets) are canned coconut milk and cream. Virgin coconut oil is usually sold in bottles. Packaged young coconut meat—shredded, flaked, toasted, sweetened, unsweetened—is available in cans or plastic bags and comes in dried, moist, or frozen form. Mature coconut meat is most often sold in desiccated or grated form but you can now find larger pieces like flakes or strips. As you shop around, see if you can find fresh or frozen variants . . . the recipes will be all the tastier with fresh ingredients!

There also are many prepared products like candies, dessert items, and sweetened coconuts ingredients. It is important to emphasize, however, that for cooking these recipes, you want ingredients that have *no added sugar*. Do not confuse sweetened cream of coconut, used mainly for desserts and mixed drinks, with unsweetened coconut milk or cream.

Storage and Serving. Coconut products in plastic bags can be stored up to six months. Coconut products in cans can be stored at room temperature up to eighteen months while unopened. Once opened, refrigerate or freeze the cans immediately. Unopened mature coconuts with a brown husk can be stored at room temperature for up to six months depending on the degree of ripeness. One medium-sized coconut will yield three to four cups of grated meat which, when stored in a well-sealed container, can be refrigerated up to four days or frozen up to six months.

With your newly acquired coconut products, try the simple-to-follow recipes below to add delicious and healthy meals to your dinner table. Enrich your own breakfast, lunch, and dinner menus. Serve the more complicated gourmet dishes for special dinners or on a special date (all the pleasures of a stimulating, exotic menu without the after-meal bloating!). As you might have guessed, the meals are perfect for beach parties or luaus, too. The recipes are diverse and creative—and some are classics given a tropical twist with the coconut. All in all, they're divinely delicious and scrumptiously sinless!

Tips: Using Virgin Coconut Oil in your Food on Daily Basis

In the previous Chapter, you learned about selecting virgin coconut oil for your skin. What next? There's more to learn about virgin coconut oil when you begin to use it in your food: liquid in temperatures below 25 °C (around 76°F), or solid at temperatures below.

Re-Liquefying Coco Butter. If your recipe calls for coconut oil in its liquid form and your oil has solidified from cooler temperatures, just warm up the coco butter till it becomes liquid again. You can try

- ➢ placing the bottle or jar of coco butter under warm running water. Or fill the sink about 1/3 of the way with warm water and leave the closed container to soak for a few minutes; or

> placing the bottle or jar in a dish or saucepan of warm water (make sure the water is not boiling as it may warp or destroy the coconut oil's container).

Do not microwave the oil because it may splatter as it liquefies.
You never need to worry about having your oil go from its solid to liquid to solid physical states. This transformation doesn't harm the oil . . . it's just the oil's natural chemistry. This is good to keep in mind when refrigerating dishes cooked with coconut oil. You will need to re-heat them before serving. For salads that you prepare and store in the refrigerator, wait till right before serving before pouring any coconut-based dressing.

Using Your Virgin Coconut Oil in Food. I suggest you first start using it as-is, for as many different things as possible: for frying, in baking, or poured over your salad. Then, experiment! To the oil, add fresh herbs, maybe pepper, and a touch of salt. Try other flavor infusions in the oil: add large chunks of ginger and garlic. Be bold . . . spice it up with fresh chili peppers or put stalks of crushed basil, thyme, and tarragon. Squeeze in lemon or lime. Or go really tropical and add *calamansi* (a tropical citrus fruit that tastes like a lemon-lime hybrid) with sea salt and mint. Afterwards, find a place in your home with the most tropical room temperature (>25C or 76°F) and keep the infused oil soaking for at least three days. Try these infused oils in your salads or when cooking.

As you begin to use virgin coconut oil in your food, I recommend that you start with small amounts of the oil. This helps you figure out how much of its taste you like in your food. This slow increase in coconut oil use also ensures that the coconut oil doesn't become a laxative for you (unless this is the effect you want). Should the laxative effect occur, lessen the amount of the oil and let your body get used to it. Those who have problems with regularity will welcome this change. But go slowly and let nature take its course.

Besides frying, baking, and cooking, or using virgin coconut oil as a salad dressing, how else can you use this oil in your food? Countless ways. For example, dip your bread in it or add it to rice. Virgin coconut oil adds a lovely, nutty taste to rice so you may wish to select an oil with a more robust coconut scent (when adding the oil after steaming). For frying rice, you may prefer a coconut oil with a slightly less coconut scent. As with any new food ingredient, experiment. You'll begin to discover your preferences for which herbs or condiments go best with virgin coconut oil, or which coconut oil you enjoy most in certain dishes.

As a general rule, you want to substitute virgin coconut oil for other oils in your own recipes, too. Remember, the dietary recommendations of the Rx: Coconuts Lifestyle: (a) *substitute* coconut oil for regular oils and fats; (b) *substitute* coconut flour for your flour needs, and coconut milk or cream for your butter needs; and (c) *selectively include* other coconut foods and coconut water.

As a special treat, following are some tasty coconut-based recipes to help you get started.

Tasty Coconut Recipes

These wonderful, flavorful recipes have been specially created by innovative master chefs, Lisa Alvendia, Beth Romualdez, Myrna Segismundo, and Pia Castillo-Lim—and their recipes reflect their diverse, international backgrounds. Besides virgin coconut oil, these recipes introduce you to other products made from coconut meat: *young and soft* (from the green nut, called *buko* in the Philippines) and *mature and firm* (from coconuts with the brown husk). Another product, not readily available outside the tropics, is coconut meat that's right in the middle—neither too soft nor too firm. The next time you're in the tropics, do treat yourself to this delicacy. Until your next island get-away, however, these tasty recipes can bring the tropics to you. Try them out today to get started on the healthful—and delicious—benefits of the coconut.

Appetizer or Cocktail Food
  Coconut Crab Dip
  Coco Vegetable Quiche

Beverages
  Coco Coffee
  Coco-mint Punch
  Cocoholic Smoothie

Rx: Coconuts! (The Perfect Health Nut)

Bread
>Pandesal Dinner Roll with 5 % Coconut Flour

Breakfast
>Eggs Benedict Casserole
>Torta de Chorizo with Coconut Cream

Dessert
>Coconut Milk Ice Cream
>Fresh Fruits with Coconut Syrup
>Coffee and Coco Jelly with Coco Cream

Meat
>Beef, Mutton, or Veal Curry
>Red Hot Beef with Apples

Pasta
>Linguini with Coconut-brandy Sauce
>Fried Noodles on Portobello Mushroom

Salad
>Grilled Eggplant Salad with Garlic-Coco Milk
>Fiesta Garden Salad with Fresh Fruits, served with Oriental Coco Dressing

Seafood
>Orange-herbed Salmon Steak with Roquefort-coconut Sauce
>Poached Fish in Turmeric Sauce
>Pan-grilled Prawns with Lemon in Coconut Milk

Snacks
>Brownies with 25%Coconut Flour
>Coco-Pineapple Macaroons

Soup
>Chicken in Coconut Soup
>Coconut Tuna Chowder

APPETIZER OR COCKTAIL FOOD

COCONUT CRAB DIP

Lisa Alvendia

2 cups sour cream
1 cup chopped fresh coconut meat or 1 cup chopped desiccated coconut
2 teaspoons curry powder
250 grams cooked flaked crab meat
1 tablespoon Dijon mustard
4 scallions, thinly sliced
¼ teaspoon cracked black peppercorns
1 teaspoon salt
¾ teaspoon chili pepper flakes or ¼ tsp cayenne pepper

Mix all ingredients in a bowl. Chill at least 3 hours before serving. Serve dip with melba toast, thin crackers or potato chips. Garnish with a sprig of fresh parsley.

Yield: 8 persons
Nutrient Content Per Serving: 86 calories; 3 grams carbohydrates; 4 grams protein; 6 grams fat

Coco-Vegetable Quiche

Lisa Alvendia

8 eggs
1 ½ cups fresh skim milk
1 cup coconut milk
1 teaspoon dijon mustard
½ teaspoon salt
¼ teaspoon pepper
1 cup grated Gruyere cheese
1 cup grated cheddar cheese or Monterey Jack

8-10 slices pandesal bread or any French bread, sliced thinly about ½"
Vegetables: slices of zucchini, spinach, button mushroom and tomatoes.

In a mixing bowl, beat and whisk eggs thoroughly and blend in skim milk. Stir well; add coconut milk, mustard and seasonings.

Line a greased rectangular pyrex dish with butter and the slices of bread as base. Sprinkle cheese over bread and make a layer of tossed-in vegetables: zucchini, mushroom and spinach. Pour the milk-egg mixture and sprinkle cheese again, then make another layer of vegetables. Top with slices of tomatoes and cheese. Bake in oven temperature 350°F for an hour or until quiche is done. Cut into desired size of square pieces and serve hot.

Yield: 8-10 persons

BEVERAGES

COCO COFFEE

Lisa Alvendia

> 1/2 cup half-and-half
> 1/2 cup **coconut milk**
> 2 cups brewed hot coffee
> 3 tablespoons coffee liqueur
> Whipped cream
> Artificial sweetener as desired

Bring milk and coffee to a boil in a large saucepan over medium heat, stirring constantly. Remove from heat. Add coffee liqueur and coconut milk. Stir well. Pour into cup. Top with sweetened whipped cream (as desired).

Yield: 4 servings
Size per serving: 3/4 cup
Nutrient Content Per Serving: 161 calories; 10 grams carbohydrates; 2 grams protein; 11 grams fat

COCO-MINT PUNCH

Lisa Alvendia

> *1cup* **coconut milk**
> *1teaspoon fresh mint, very finely chopped*
> *1/2 teaspoon salt*
> *1/4 teaspoon ground pepper*
> *1 1/2 cups club soda*

Mix coconut milk, mint, salt, and pepper. Stir well.
Pour into a pitcher.
Add club soda, stirring constantly.
Add several ice cubes and mix again.
Serve in tall glasses.
Yield: 3 servings
Size per serving: 7 ounces
Nutrient Content Per Serving: 195 calories; 8 grams carbohydrates; 1.5 grams protein; 20 grams fat

COCOHOLIC SMOOTHIE

Lisa Alvendia

> *2 cups* **coconut ice cream,** *see page 103 (if not available, use vanilla ice cream)*
> *2 bananas, sliced*
> *1/4 cup* **coconut milk**
> *1/4 cup fresh milk*
> *Ice cubes*
> *Optional: dash of vodka or* **toasted coconut**

Process the first four ingredients in a blender.
Add enough ice cubes to bring mixture to 4 1/2-cup level.

Process until smooth, stopping once to scrape the sides.
Serve immediately in a tall beverage glass with a straw.
Optional: ad a dash of vodka and/or top with toasted coconut.
Garnish with a sprig of mint leaves.
Yield: 4 servings
Size per serving: 6 ounces
Nutrient Content Per Serving: 190 calories; 22 grams carbohydrates; 3 grams protein; 10 grams fat

BREAD

PANDESAL WITH 5% COCOFLOUR

3Rs on Coconut Flour by the Philippine Coconut Authority

Bread flour 1900 grams	Olympial soft (bread improver) 20 grams
Sugar 300 grams	Shortening 50 grams
Powdered milk 100 grams	Egg 1 piece
Coconut Flour 100 grams	Bread crumbs
Cold water 1000 grams	Iodized salt 30 grams
Instant yeast 40 grams	Butter 50 grams
Vanilla 10 grams	

Weigh all ingredients. Mix bread flour, coconut flour, powdered milk, salt, yeast, butter and egg. Put in dough mixer. Then add the shortening gradually. Dissolve sugar in cold water and pour gradually to the dry ingredients. Add vanilla. Mix the dough for 18 minutes or until smooth and elastic. Cut into four equal parts. Rest for 30 minutes. Roll evenly then roll in bread crumbs. Cut into desired size, 1 ½ inch (cut in upper right portion). Put in a tray. Proof for 45 minutes.

Yield 112 pieces 30 grams each.

BREAKFAST FOOD

EGGS BENEDICT CASSEROLE

Lisa Alvendia

1 ½ cup cut-up smoked ham (fully cooked)
6 eggs, poached
¼ teaspoon pepper
½ cup crushed cornflakes (cereal)
¼ cup butter, melted
3 pieces English Muffins (sliced into halves)

Line a greased rectangular baking dish with 6 pieces. English muffins. Sprinkle ham on top of the bread and top each with poached eggs. Sprinkle eggs with pepper. Pour Coco-Mornay sauce on top.

COCO-MORNAY SAUCE:

¼ cup butter or coconut butter
¼ cup all purpose flour
2 cups fresh milk
½ cup **coconut milk**
½ teaspoon salt
¼ teaspoon nutmeg
1 ½ cup shredded Gruyere or Swiss cheese
½ cup grated Parmesan cheese

Heat butter in a saucepan over low heat. Blend in flour and seasonings. Cook over low heat, stirring constantly with a wire whisk until smooth and bubbly. Stir in milk and heat to boiling, stirring constantly. Add cheese and stir until cheese is melted and mixture is smooth.

How to poach eggs: For easy results, there is a microwaveable 2-egg poacher: Break one fresh egg into each hole of the poacher. Pour one tsp water on top of each raw egg. Cover the poacher tightly and heat one minute in the microwave. Spoon out the poached eggs and continue to do 4 more eggs.

Arrange the poached eggs, one on top of each English muffin half; pour the Mornay sauce on top of the eggs. Toss the crushed cornflakes with melted butter, arrange around each egg and the rest around the edges of the pyrex dish. Bake 350°F for 30 minutes until hot and bubbly. Sprinkle each egg with paprika. Serve Hot.

Yield: 6 pieces

TORTA DE CHORIZO WITH COCONUT CREAM

Lisa Alvendia

> 1 tablespoon **virgin coconut oil**
> 1 small onion, chopped
> Fresh herbs, each 1 tablespoon: mint, parsley, cilantro and basil
> 1/2 cup sliced chorizo or kielbasa, half-moon
> 4 whole eggs, separate yolk and white
> 1/2 teaspoon salt
> 1/4 teaspoon pepper
> 1/4 cup **coconut cream**

Sauté onions and chorizos in oil. Immediately add the fresh herbs. Stir well over medium heat. Cook for 2 minutes then set aside.

Beat egg yolks at high speed with an electric mixer (or with a wire whisk) until thick and lemon-colored. Mix in the coconut cream. Add salt and pepper.

In a separate bowl, beat egg whites vigorously until stiff but not dry. Fold into the egg yolk mixture gently. Add the chorizo mixture and fold gently.

Melt butter in a skillet and pour the egg-chorizo mixture.

Spread egg-chorizo mixture to make a torta.

Cook slowly until the egg is firm and lightly-browned on the bottom. Repeat.

Makes 2 whole tortas. Slice the tortas in half. Stack 1/2 tortas on top of each other, pancake style.

Pour the torta sauce over the center and along the sides of the torta.

TORTA SAUCE:

1 tablespoon coconut oil
1 can plum tomatoes
3 tablespoon catsup
2 medium onions, thin half-moon slices
1 tbsp. chili powder
Salt and pepper to taste

Boil all ingredients together and remove from heat. Serve hot.

Yield: 4 slices
Size per serving: 1 slice
Nutrient Content Per Serving: 425 calories; 8 grams carbohydrates; 12 grams protein; 40 grams fat

DESSERTS

COFFEE AND COCO JELLY WITH COCO CREAM

Myrna Segismundo

Coffee jelly

1/2 cup unflavored gelatin
2 1/2 cups water
1 tablespoon instant coffee, decaffeinated
1/2 cup white sugar (adjust as desired)
Coco jelly
1/2 cup unflavored gelatin
*2 1/2 cups **coconut milk** (fresh or canned)*
1/2 cup white sugar (adjust as desired)

For each jelly: In a preheated sauce pan, combine gelatin and water and simmer over low heat. Add coffee and sugar and stir to dissolve in liquid. Pour into a bowl and chill to set. Cut into cubes before serving.

Rx: Coconuts! (The Perfect Health Nut)

Coco cream

2 cups coconut milk
2 tablespoon white sugar

In a preheated pan, simmer coconut milk and sugar, stirring occasionally to avoid lumps. Strain coconut cream and allow to cool.

Combine coffee and coco jelly in individual serving glasses and lace with **coco cream**. Serve chilled.

Yield: 20 servings
Size per serving: 1/2 cup
Nutrient Content Per Serving: 197 calories; 24 grams carbohydrates; 1 grams protein; 12 grams fat

COCONUT MILK ICE CREAM

Pia Castillo-Lim

*2 cups **coconut milk** extracted from freshly grated **coconut** (1 large coconut) or 2 cups of canned **coconut milk***
1/2 cup glutinous rice flour (available at Asian stores)
2 tablespoon cane sugar, or brown sugar to taste
50 grams of moscovado sugar (optional)
1 pandan or screwpine leaf for flavor or 1 tsp anisette or 1 tsp almond extract

Extract the milk from one coconut adding warm water to the first extraction to make 2 cups.

Mix the milk with the rice flour and the 2 tbsp. white sugar and cook over low heat until the mixture thickens (about 15 to 20 minutes). It is important to keep the heat low so that the milk does not curdle. Add a piece of pandan or screwpine for flavoring.

Cool the mixture once it thickens.

When cooled, put the mixture in your ice creamer and churn the mixture.

When it is halfway done, add the 50 grams of moscovado sugar. This will add to the sweetness and give you bits of sugar to bite into when done. Freeze the ice cream.

This is the basic ice cream. You may choose to add other complementary flavors such as:

2 tablespoon of toasted white sesame seeds, or
1/2 cup of sweetened nata de coco (available at Asian stores), or
1/2 cup of sweetened jackfruit (available at Asian stores).

Note: Because this creamy ice cream contains only coconut milk and glutinous rice flour, you can and guiltlessly enjoy it, even if you are lactose intolerant.

Yield: 8 individual medium-sized scoops
Size per serving: 1 medium-sized scoop
Nutrient Content Per Serving: 205 calories; 17 grams carbohydrates; 2 grams protein; 15 grams fat

FRESH FRUITS WITH COCONUT SYRUP

Lisa Alvendia

> *1 cup honeydew melon balls*
> *1 cup cantaloupe balls*
> *1 cup thick* **coconut cream**
> *1/2 cup diced pineapple*
> *1/2 cup diced papaya*
> *3 cups sugar*
> *2 cups crushed ice*
> *2 cups water*
> *Few drops jasmine extract (optional)*

Scoop honeydew and cantaloupe into balls. Dice pineapple and papaya; set aside.

In a saucepan, boil sugar and water until syrupy. Strain and cool. Add jasmine extract, if desired. Arrange fruit in a serving bowl or in individual serving dishes and top with syrup, coconut milk, and crushed ice.

Yield: 8 cups
Size per serving: 1/2 cup
Nutrient content per Serving: 421 calories, 84 grams carbohydrates, 1.6 grams protein, 10 grams fats

MEAT

RED-HOT BEEF WITH APPLES

Lisa Alvendia

1 pound lean beef (sirloin), sliced into 1"thin strips
2 tablespoons extra virgin olive oil
2 tablespoons flour
1 teaspoon chili pepper flakes or 1 piece jalapeno pepper, chopped and seeded
2 tablespoons soy sauce
2 tablespoons curry powder
1 tablespoons **virgin coconut oil**
1 cubed beef beef bouillon (dissolved in ½ cup water and 1 cup **coconut milk**
½ chopped onions
1 tablespoon lemon juice
2 tablespoons honey or apple jelly
½ cup shredded **fresh coconut meat** *or ½ cup shredded,* **desiccated coconut**
1 large apple, cored, seeded and sliced into 8 wedges
2 scallions, sliced thinly on the diagonal
Boiled Jasmine scented or Thai fragrant rice
Selection of condiments to serve: raisins, chopped hard-boiled egg, chopped cashew nuts, sliced pineapple chunks or any chutney.

Marinade beef slices with chili pepper flakes, curry powder and soy sauce; cover and let stand for an hour. Sprinkle flour over the beef and toss well to coat evenly. In a large skillet, fry the beef strips in oil until lightly browned. Remove from heat and set aside. In the same skillet, sauté onions with coconut oil until translucent and put the beef back. Add coco-bouillon milk mixture and simmer for a half hour or until beef is tender. Add the coconut meat, apples, lemon juice and honey. Simmer for 10 minutes and add scallions.

Serve the dish in the center of a ring mold of boiled rice with a selection of individual condiments on the side.

Yield: 4 persons
Nutrient Content Per Serving: 292 calories; 16 grams carbohydrates; 30 grams protein; 14 grams fat

BEEF, MUTTON, OR VEAL CURRY

Lisa Alvendia

> *2 pounds beef, mutton or veal*
> *1/2 teaspoon coriander*
> *2 tablespoons mustard oil*
> *1 1/2 teaspoons salt*
> *4 teaspoons minced onion*
> *1 cup thick* **coconut cream**
> *1 teaspoon ground chili*
> *Juice of one lemon*
> *1/2 teaspoon ground ginger*
> *4 tablespoons ghee (clarified butter)*
> *1/4 teaspoon minced garlic*
> *1 teaspoon turmeric*

Cut meat into 3/4-inch thick slices. Lay in a marinade of oil, spices, salt, coconut cream and lemon juice.

After 20 minutes, cut meat into squares; return to mixture and soak for an hour.

Pierce meat with metal skewers and roast over red, hot coals, basting with clarified butter.

Serve hot.

Note: The marinade may be cooked in a little butter, flour, and water to make a rich gravy for the beef.

Yield: 8 servings
Size per serving: 100 grams (3 cubes)
Nutrient Content per Serving: 324 calories, 2 grams carbohydrates, 30 grams protein, 22 grams fats

PASTA

FRIED NOODLES ON PORTOBELLO MUSHROOMS

Lisa Alvendia

> *4 large portobello mushrooms (stems removed)*
> *2 tablespoons extra virgin olive oil*
> *1 teaspoon salt*
> *½ teaspoon freshly ground pepper*
> *3 tablespoons virgin coconut oil*
> *2 cloves garlic, macerated*
> *½ teaspoon chili pepper flakes*
> *Mixture of 1 tablespoon tomato paste*
> > *1 teaspoon paprka*
> > *2 tablespoons light soy sauce*
> > *½ cup chicken broth*
>
> *Egg noodles (4-6 ounces cooked as directed by box)*
> *½ cup red pepper, sliced into strips*
> *3-4 pieces. cabbage leaves, sliced thinly*
> *Salt and pepper to taste*
> *Garnishing: A handful of flat-leaf parsley or cilantro leaves.*

Heat oven to 350°F. Arrange the mushrooms in a baking pan. Brush the mushrooms with olive oil and sprinkle the open caps with salt and pepper. Bake for 20-25 minutes. Remove from pan and set aside. Heat coconut oil in a wok and sauté garlic. Add cabbage and red pepper, chili flakes and the seasoning mixture. Simmer for 3-5 minutes and add noodles. Stir in the whole mixture and mix noodles thoroughly. Taste and adjust seasoning.

To serve: Put each portobello mushroom on a warm dinner plate, gills upward. Divide the noodles with vegetables mixture among the 4 mushrooms. Garnish with parsley or cilantro.

Yield: 4 persons

LINGUINI PAGODA WITH COCONUT-BRANDY SAUCE

Lisa Alvendia

Linguini (8-ounce box), cooked as directed on box
3 tablespoons butter
2 cloves garlic, macerated
1 medium onion, chopped
2 tablespoons fresh parsley, chopped
1/2 teaspoon each of nutmeg, cumin, pepper, salt
1/2 cup brandy
2 cups **coconut milk**
1 cup skim milk
2 cups heavy cream, whipped
1/4 cup grated parmesan cheese
½ cup button mushrooms, sliced
1 cup chopped fresh spinach

In a skillet, sauté garlic and onions in butter over medium heat.

Stir in parsley and other spices. Set aside. In a separate bowl, mix coconut milk, cream, and skim milk. Heat to boiling, and reduce heat, simmering uncovered. Add the brandy and parmesan cheese, stirring frequently until thickened. In a large container, mix the cooked linguini over the sauce and blend in well. Sauté button mushrooms with butter and add spinach until done.

To serve: Put a small quantity of noodles in the middle of each plate. Top this with a spoonful of mushroom-spinach mixture and finally more noodles on to of it to make it like a pile of layered noodles. Top with another layer of mushrooms and garnish with a flat-leaf parsley. Serve immediately

Yield: 6 servings
Size per serving: 1 cup
Nutrient Content Per Serving: 302 calories; 20 grams carbohydrates; 6 grams protein; 24 grams fat

SALAD

GRILLED EGGPLANT SALAD WITH GARLIC-COCO MILK

Myrna Segismundo

> *2 pounds eggplant, grilled, peeled and coarsely chopped*
> *1 cup **coconut milk** (fresh or canned)*
> *3-4 cloves garlic, minced*
> *2 tablespoons lemon juice*
> *Freshly ground black peppercorns, to taste*
> *Salt, to taste*

Combine grilled eggplant and coconut milk. Add minced garlic and toss well. Season with lemon juice, black peppercorns, and salt. Chill before serving.

Yield: 12 servings
Size per serving: 1 piece
Nutrient Content Per Serving: 66 calories; 6 grams carbohydrates; 1 gram protein; 5 grams fat

FIESTA GARDEN SALAD WITH FRESH FRUITS, SERVED WITH ORIENTAL COCO DRESSING

Beth Romualdez

> 1 head each of 3 different kinds of lettuce, torn into bite-size pieces, or about 1 1/2 cup per variety of lettuce
> 1/2 cup of each of the following fruits, sliced into 1/2-inch wedges: cantaloupe, strawberries, jackfruit, chico, mango, seedless grapes (preferably red or black), apple (cored, seeded, and immersed in lemon juice to avoid discoloration)
> 1 pack (100 grams) crystal noodles
> Cooking oil for frying

Immerse cut-up lettuce leaves in ice water for 1/2 hour. Drain and pat dry with paper towel. Set aside.

Refrigerate sliced fruits separately and set aside.

Fry the crystal noodles in very hot oil (450°F). Remove from heat as soon as they puff. Place in a colander and pat dry with paper towel. Set aside.

Meanwhile, make your oriental coco dressing.

ORIENTAL COCO DRESSING

> 1 cup **cocojam**
> 1/4 cup **coconut milk**
> 3 tablespoons balsamic vinegar
> 1/2 cup water
> 1 tablespoon soy sauce
> 1 1/2 teaspoon peppercorns, crushed
> 1/2 teaspoon salt
> 1 tablespoon extra virgin coconut oil
> 1/2 teaspoon sesame oil
> 1 1/2 tablespoon toasted sesame seeds

In a medium mixing bowl, combine with a wire whisk all ingredients except sesame seeds. This serves as your salad dressing. Set aside.

In a large ceramic platter, place salad greens. Arrange assorted fruits, making colorful toppings. Crush toasted noodles and sprinkle on the salad. Top with sesame seeds. Drizzle with the oriental coco dressing. Serve extra salad dressing on the side.

Yield: 6-8 servings
Size per serving: 1/2 cup
Nutrient Content per Serving: 125 calories, 22 grams carbohydrates, 1 grams protein, 4 grams fats

SEAFOOD

ORANGE-HERBED SALMON STEAK WITH ROQUEFORT COCONUT SAUCE

Lisa Alvendia

Marinade: 2 tablespoons grated orange zest
1/4 cup orange juice freshly squeezed
3 tablespoons fresh lemon juice
3 tablespoons fresh dill weed, chopped
3 tablespoons fresh cilantro leaves, chopped 1/2 inch
2 tablespoons virgin **coconut oil**
1/2 teaspoon salt combined with 1/4 teaspoon cracked peppercorns
8 salmon steaks, 1 1/2 inches thick
3 tablespoons olive oil

Combine first 5 ingredients together in a large shallow dish. Set aside.

Rub the salt and pepper mixture into the salmon steaks and add coconut oil on both sides.

Arrange salmon steaks and rub the marinade on both sides. Cover and chill for 2-3 hours. Then remove salmon steaks from marinade.

Heat a skillet with oil and over a very high temperature (475°F), pan-sear salmon steaks for 7-10 minutes on each side, leaving the outside cooked and the inside medium rare. Set aside.

Serve with Roquefort Coconut Sauce

ROQUEFORT COCONUT SAUCE

1/4 cup Roquefort or blue cheese, crumbled and mashed
1/2 cup cream, whipped
*1/4 cup **coconut cream***
1/4 cup scallions, chopped finely
1/4 cup cilantro, or fresh parsley chopped finely
2 tablespoons fresh dill weed
1 teaspoon salt
1/2 teaspoon red pepper flakes

Blend the first three ingredients together in a mixing bowl, then add the fresh herbs and seasonings. Mix all the ingredients together.

To serve, arrange salmon steaks on a large platter.

Garnish with alternate slices of fresh cucumber and tomatoes

Serve the Roquefort-coconut sauce on top center of your salmon and around it. Garnish with alternate slices of cucumber and tomato.

Yield: 6 servings
Size per serving: 100 grams
Nutrient Content Per Serving: 253 calories; 4 grams carbohydrates; 32 grams protein; 12 grams fat

PAN-GRILLED PRAWN WITH LEMON IN COCONUT MILK

Beth Romualdez

12 medium-sized tiger prawns, deveined, shell on
Salt and pepper
*2 tablespoons **virgin coconut oil***
3 cloves garlic, peeled and chopped
1 small onion, sliced
6 pieces medium-sized kamias, seeded and sliced
*1 cup **coconut milk***
2-inch lemongrass stalk, crushed (preferably fresh, dried is a poor substitute)
2 finger chilies, seeded and sliced

2 tablespoons finely diced tomatoes

Wash the prawns well and cut open from the head down to the tail. Devein and rinse. Season with salt and pepper.

In a non-stick skillet, heat the oil and grill the prawns on both sides. Transfer the prawns to a platter.

In the same pan, add the garlic and onions. When the onions are soft, add the coconut milk and the lemongrass.

Cook while stirring, and when it starts to simmer add the kamias and the green chili.

Cook until the kamias and chili are soft and the coconut milk has reduced a bit. Season with salt.

Remove from fire, discard the lemongrass, and put the coconut sauce in a blender and puree.

To serve: ladle about half of the coconut sauce in each plate, top with the grilled prawns and sprinkle the tomatoes on the sauce and around the plate.

Yield: 6 servings
Size per serving: 2 pieces
Nutrient Content per Serving: 252 calories, 5 grams carbohydrates, 26 grams protein, 16 grams fats

POACHED FISH IN TURMERIC SAUCE

Pia Castillo Lim
    2 pounds whole meaty fish like tuna, milkfish, or seabass, descaled and seasoned with salt and pepper
    1 whole piece of fresh turmeric, sliced, or 1 tbsp. ground turmeric powder
    *3 large **coconuts, grated and milk extracted** or 2 cups canned or UHT packed **coconut milk***
    1-4 finger chilies, to taste
    Fish sauce and pepper, to taste

Grate the coconut and extract the pure milk. This is coconut cream.

Set aside. (To save time, you may buy coconut milk in UHT packs. You will need about 2 cups of pure coconut milk for this recipe.)

Add 1/2 cup of warm water to the squeezed coconut and extract as much coconut milk as you can from this. Simmer the second extraction till it thickens.

Add the sliced turmeric and the finger chilies.

Bring the mixture to a simmer and allow the sauce to evaporate until thickened.

When almost ready to eat, drop the seasoned fish in and poach it till half cooked (about 10 minutes per inch when you measure the fish on its fattest side). Still in simmer, add the coconut cream to thicken the sauce (5 to 10 minutes).

Season with fish sauce and pepper.

Note: This method of poaching fish in coconut milk can be used also for poaching chicken, pork, beef, and an assortment of vegetables. This recipe uses turmeric as its main flavoring, but you can also replace it with any of the following: Kaffir lime leaves, basil, lemongrass, ordinary ginger, or fermented fish paste (bagoong). Use your creativity to blend these ingredients to come up with an interesting and tasty dish.

Yield: 8 servings
Size per serving: 1 slice
Nutrient Content Per Serving: 351 calories; 4 grams carbohydrates; 32 grams protein; 24 grams fat

SNACKS

BROWNIES (25% cocoflour)

From 3Rs on Coconut Flour by the Philippine Coconut Authority

1 cup butter
1/4 cup **coconut flour**
2 cups sugar
1/2 tablespoon Baking soda
6 eggs

Rx: Coconuts! (The Perfect Health Nut)

1/2 teaspoon salt
2 teaspoon Vanilla
1 cup cocoa
3/4 cup all-purpose flour
1 cup chopped nuts

Cream butter then add sugar gradually.

Add eggs one at a time mixing well after each addition. Mix in vanilla.

In another bowl, mix together all-purpose flour, coconut flour, baking soda, salt and cocoa.

Add flour mixture to the creamed butter and mix until well blended. Add 1/8 cup of chopped nuts.

Pour into greased pan. Smoothen surface with a rubber scraper and top with the remaining nuts.

Bake for 350°F for about 15 minutes.

Yield: 15-20 pieces
Size per serving: 1 piece
Nutrient Content Per Serving: 195 calories; 30 grams carbohydrates; 5 grams protein; 7 grams fat

COCO-PINEAPPLE MACAROONS

Myrna Segismundo

2 eggs
1/2 cup unsalted butter, melted
1/4 cup white sugar
1 cup condensed milk
1 cup crushed pineapple (canned or fresh), drained
1 1/2 cups **desiccated coconut**
20 (1 1/4-inch) baking cups

Preheat oven at 300°F. In a mixing bowl, whisk eggs, butter, and sugar. Add condensed milk and stir to blend well.

Fold in crushed pineapple and desiccated coconut and spoon mixture into individual paper baking cups placed in 1 1/4 inch round muffin pans.

Bake for 20 minutes or until macaroons are golden brown. Allow to cool before serving.

Yield: 20 pieces
Size per serving: 1 pc (1 1/4")
Nutrient Content Per Serving: 117 calories; 18 grams carbohydrates; 2 grams protein; 4 grams fat

SOUPS

CHICKEN IN COCONUT SOUP
Ideally served in fresh buko (young coconut) shells

Lisa Alvendia

> 2 tablespoons virgin coconut oil
> 1 cup fresh chicken breast fillet sliced into 1" strips
> 2 cups **coconut cream**
> 2 cups **chicken broth**
> 1 thumb-size ginger, sliced and pounded slightly
> 1/2 cup sour cream
> 1/4 cup heavy cream, whipped
> 2 tablespoons salted shrimp sauce (optional)
> 1 tablespoon fresh lemon juice
> 1 teaspoon chilli powder
> 3 tablespoon coriander leaves, fresh and sliced 1/2" thick
> Salt and pepper to taste

Saute chicken breast in coconut oil over medium heat. Add ginger, coconut cream and chicken broth. Cook until chicken breast is tender. Season with salted-shrimp sauce; add in sour cream and heavy cream. Stir well and let boil for 5 minutes; remove from heat and set aside.

For garnishing: Serve coconut soup in individual buko shells which will serve as your soup bowl. Place in a wooden plate or any desired plate soup underline. Line with a cut-out round of a fresh banana leaf. If not available, you can line it with a laced-paper cake doily.

Yield: 8-10 servings
Size per serving: 6 ounces
Nutrient Content Per Serving: 195 calories; 4 grams carbohydrates; 7 grams protein; 18 grams fat

COCONUT TUNA CHOWDER

Lisa Alvendia

> 3 seven-ounce cans tuna
> 1/4 cup **virgin coconut oil**
> 1/3 cup flaked **coconut**
> 3 tablespoons flour
> 2 cups chicken broth
> 1/2 teaspoon salt
> 2 medium onions, sliced
> 1/2 teaspoon curry powder
> 3 ribs celery, sliced
> 1/4 teaspoon ginger threads
> 2 cloves
> 2 cups milk
> 6 peppercorns
> 1 cup light cream
> 1 bay leaf

Drain and flake tuna. Set aside.

Toast coconut flakes in oven for 5 to 8 minutes or until golden brown; set aside.

Boil together broth, onion, celery, cloves, bay leaf, and peppercorn and simmer 30 minutes; set aside.

In a large skillet put in oil and blend with flour, salt, curry powder, and ginger. Gradually blend in milk.
Bring to a boil, stirring constantly until sauce thickens.
Add broth mixture, discarding bay leaf.
Add tuna and light cream. Heat thoroughly.
Top each serving with toasted coconut flakes.

Yield: 10 servings
Size per serving: 6 ounces
Nutrient Content Per Serving: 322 calories; 10 grams carbohydrates; 21 grams protein; 22 grams fat

The Chefs . . .

Beth Romualdez: On both sides of the Pacific Ocean, Beth is a teacher, food, and beverage consultant on Italian, Balinese, Indonesia, Thai, Lao, Vietnamese, Asian fusion, and Philippine cuisine. Anvil Philippines published Beth's *Cooking Lessons* in 2004. The gorgeous simplicity of her recipes, together with detailed cooking tips and how to's, made her book an instant hit, and it continues to be ranked in the top ten of best-selling books in the Philippines.

Myrna Segismundo is the multi-awarded managing director of *9501*, the exclusive executive lounge for ABS-CBN. She is the author of *The Party Cookbook,* which features stunning recipes from sixteen years of culinary work at the prestigious Sign of the Anvil Executive Lounge. The book reflects her finesse and organizational skills from years of "catering to the stars".

Lisa Alvendia is a sought after theme party consultant. Her specialty is the planning and execution of beautifully crafted parties. Her 2001 book, *Creative Catering and Entertaining,* by Anvil, Philippines has become a must-have for homemakers and professionals looking to create the perfect party.

Pia Lim-Castillo: With a background in agricultural economics and food anthropology, Pia regularly writes and gives lectures on Philippine food traditions, history, and indigenous ingredients. A member of Slow Food International, as President of its Manila Convivium, she leads a group that celebrates quality of life, dining as a source of pleasure, and agricultural biodiversity.

Imelda C. Peralta, Chief Dietician at the Makati Medical Center Hospital, brought her academic expertise in calculating the improved calories of coconut oil in these recipes.

Beyond the Recipes: Sustained Health for Life

These recipes showcase the coconut in fine cooking and give you a teensy taste of just how widely it can be used. My daughter's cook (who's been with our family for over thirty years) has been experimenting herself and has made some delicious, coconut-inspired concoctions: a salad dressing that mixes spicy Thai chili sauce with coconut oil (drizzled over a friend's grapefruit salad recipe); tofu, bean sprouts and mixed vegetables sautéed in a light coconut gravy; and their household's new favorite, vegetable lasagna with pumpkin slivers as "pasta," fresh cheddar and mozzarella cheese and a lovely coconut-and-tomato sauce. She's even begun on desserts, her current specialty being a light, fluffy cheesecake made with egg, coconut cream, low-fat milk, Splenda®, and light cream cheese—with blueberries or strawberries on the side. Even better, this forward-thinking cook has started to make sure that all her food at home is prepared with coconut oil for the benefit of her daughter and infant granddaughter. As you can see, substituting coconut oil in food is an affordable, easy, and tasty way to encourage health in your family and in your staff.

In Chapter 1 you learned that coconut oil has slightly less calories per gram than all other fats. So when you bake and cook, substituting the lower calories-per-gram coconut oil for the higher-calorie oils and butter or margarine will automatically reduce the number of calories of the fat you eat by 2.5%. You also learned that the coconut's medium-chain fatty acids break down faster, improve your metabolism, and make you feel full, faster. The bottom line is: whenever you substitute coconut oil for other oils in your food preparation, you can eat tasty food that is less likely to go right to your belly and hips!

In other Chapters, you learned how ingesting coconut oil gives you other benefits: it is cholesterol and trans fat free, and doesn't produce blood clotters. All this helps to keep your heart healthier. Virgin coconut oil can also keep internal and external infections at bay and may even help prevent cancer. You learned how beneficial coconut oil and its derivatives are to the skin and how new studies are starting to show the great potential of coconut growth factors for our internal and external rejuvenation.

So after reading this book and going through all the recipes—twice—does our coconut journey end here? Hardly. Refer often to Chapter 11 (Rx 3) and the Rx: Coconuts Lifestyle . . . it's your daily "prescription" for head-to-toe, cell-to-surface health and beauty. It's a simple, easy commitment to your internal and external health and it maximizes the benefits of the coconut—its oil, food, water, and other derivatives—in as many areas of your life (and your family's) as possible.

In a nutshell, here's wishing you a lifetime of heart fitness, weight loss, protection from disease, cancer prevention and overall natural gorgeousness—all the best of this Perfect Health Nut!

Directory Of Coconut Products

Retailers: Stores That Sell Coconut Products (Australia, Europe, North America, Online)
Retailers and Suppliers of Specialized Coconut Products
Suppliers: Where to Have Your Own Coconut Products Developed or Manufactured
For questions on Philippine virgin coconut oil: *www.pca.da.gov.ph*
For questions to the author: *vmvr@qinet.net* or vmvr@aol.com

Retailers: Stores Around the World that Sell Coconut Products

AUSTRALIA

Bi-lo
www.bilo.com.au
Coles
www.coles.com.au

GIA
Payless
Woolworths
www.woolworths.com.au

CANADA

JLB
Toronto
Tel no. 416.487.2813
Whole Foods Market
Many locations in Canada, UK, and USA
301 Cornwall Road
Oakville, Ontario L6J 7Z5
905.815.0266
www.wholefoods.com

Korean Central Market
675 Bloor Street West
Toronto, ON M6G 1L3
(416) 516-3966

ONLINE

VMV Hypoallergenics www.vmvhypoallergenics.com

EUROPE AND NORTH AMERICA

UNITED KINGDOM

Banton Store
Bath
King Yip
Bristol

King Yip
Bristol

UNITED STATES OF AMERICA

ARIZONA

Western Market
4208 W. Dunlop Ave., Phoenix

Olympic Market
1130 West Guadelupe Road #5, Mesa

Rx: Coconuts! (THE PERFECT HEALTH NUT)

CALIFORNIA

Kukjae Market
2350 Junipero Serra Blvd.,
Daly City

Arirang Super Market
8281 Garden Gove Boulevard,
Garden Grove

California Market
450 South Western Avenue,
Los Angeles

Han Nam Chain
2740 West Olympic Blvd.,
Los Angeles

Assi Plaza
3525 West 8th Street, Los Angeles

Plaza Market
928 South Western Avenue,
Los Angeles

Tai Lee Supermarket
Los Angeles

Island Pacific Korean
Los Angeles

**Koreatown Galleria
Super Market**
3250 West Olympic Blvd.,
Los Angeles

Korean Food Market
332 14th Street, Oakland

Pusan Market
2370 Telegraph Avenue, Oakland

Oriental Food
9180 Kiefer Blvd., Sacramento

First Korean Market
4625 Geary Blvd., San Francisco

Geary Food Market
4324 Geary Blvd., San Francisco

Kaju Market
2000 Judah Street, San Francisco

Kukjea Market
3130 Noriega Street, San Francisco

Kyopo Market
3379 East El Camino Real, Santa Clara

Farm Market
12500 E. Slauson Avenue,
Santa Fe Springs

Han Kook Super Market
1092 East El Camino Real, Sunnyvale

South Pacific Trading Company, Inc.
22601 E, La Palma Ave., Yorba Linda

CONNECTICUT

Oriental Pantry
374 Whitney Avenue, New Haven

FLORIDA

Ada's Natural Food Supermarket
11705 S. Cleveland Avenue,
Fort Myers

Wards Supermarket, Inc.
515 N.W. 23rd Avenue, Gainesville

Seoul Market
200 North State Road 7,
Hollywood

Organic Food Center
862 N. Hwy. A-1-A, Indialantic

For Goodness Sake (Naples)
7211 Radio Road, Naples

Vinh An Market
372 NE 167th Street, North Miami Beach

Super Value Nutrition
5816 Bee Ridge Road, Sarasota

Perfect Balance Organics
9264 Cortez Blvd., Weeki Wachee

Publix
Sedans

Seoul Oriental Market
8031 W. Oakland Park Blvd., Sunrise

New Leaf Market
1235 Apalachee Parkway, Tallahassee

GEORGIA

Buford Highway Farmer's Market
5600 Buford Highway, Atlanta

Han Kuk Shik Poom
3656 Buena Vista Road, Columbus

Lotte Oriental Food
5224 Bulford Highway, Doraville

ILLINOIS

Chicago Food Market
5800 North Pulaski Highway,
Chicago

Chicago Food Corporation
3333 North Kimball Ave., Chicago

INDIANA

Kim's Oriental Grocery
8710 E. 21st Street, Indianapolis

IOWA

Wang's Asian Market
3005 100th Street, Urbandale

KENTUCKY

Oriental Supermarket
5083 Peston Highway, Louisville

LOUISIANA

Oriental Market
3324 Transcontinental Drive, Metairie

MARYLAND

Dae Jun Supper
1808 Norht Patterson
Park Avenue, Baltimore

Eastern Food Market
2030 Paul Street, Baltimore

Jin Mi Oriental Market
10800 North Rhode Island
Avenue, Beltsville

Han Ah Reum-Baltimore
800 N. Rolling Road, Catonsville

Lotte Plaza-Ellicot City
8801 Baltimore National Pike,
Ellicott City

Grand Mart-Gaithersburg
221 Muddy Branch Road,
Gaithersburg

Grand Mart-Germantown
12851 Clopper Road, Germantown

Landover Oriental Food
7535 Landover Road, Landover

Lotte Plaza—Rockville
11790 Parklawn Drive, Rockville

Lotte Plaza—Silver Spring
13625-A Georgia Avenue, Silver Spring

Dae Sung Oriental Grocery
2213 Green Spring Drive, Timonium

Nu-Kor Oriental Foods
714 M York Road, Towson

Korean Korner International Market
12207 Veirs Mill Road, Wheaton

Han Ah Reum Asian Supermarket
12015 Georgia Avenue, Wheaton

MASSACHUSETS

Chung Ge Kim's
210-D West Main Street, Ayer

Reliable Market
45 Union Square, Sumerville

MINNESOTA

Seoul Oriental Foods
1071 E. Moore Lake Drive, Fridley
Dong Yang Oriental Food & Deli

735 45th Ave. NE, Hilltop

Kim's Oriental Foods and Gifts
689 N. Snelling Ave., St. Paul
Healthkeepers Market & Wellness Center
27764 North Star Road, Winona

MISSISSIPPI

Sweet and Sour
307 Clinton Blvd. #C, Clinton

NEVADA

Asian Market
953 East Sahara Avenue #7-9, Las Vegas

NEW JERSEY

Hans Food
433 South Washington Avenue, Bergenfield
Edison Hanahreum
518-14 Oldposted Road, Edison
Fil-Am Food Mart
685 Newark Ave., Jersey City

Onion Cellar
220 Broad Avenue., Palisade

Hanahreum
321 Broad Avenue, Ridgefield

Rx: Coconuts! (The Perfect Health Nut)

NEW YORK

Food Emporium
New York City
Gristedes
New York City
Kim's Oriental Gift & Food
811 Millersport Highway, Amherst
Cortland Health Foods
64 Main Street, Cortland
Assi Plaza
131-01 39th Avenue, Flushing

Han Ah Reum Asian Mart
47-08 Grand Ave., Maspeth
Han Garam
2737 Erie Boulevard, Syracuse
Hanahreum
29-02 Union Street, Flushing
Sung Oriental Grocery & Gift
850 Niagara Falls Blvd., Tonawanda

NORTH CAROLINA

Lotte Oriental Food
4211 South Boulevard, Charlotte
Lotte Oriental Food
6311-C Glenwood Avenue, Raleigh

Han Mi Oriental Food Market
128 North Main Street, Spring Lake

OHIO

Toul Bo Oriental Market
5046 North High Street, Columbus
Dong Yang Oriental Food & Gifts
500 West Main Street, Fairborn

Dong Won Oriental Market
4271 Mayfield Road, South Euclid

OKLAHOMA

Korean Oriental Foods
4520 South Bryant Road, Oklahoma City

OREGON

Go Bu Gi Market
4520 Southwest 110th Avenue, Beaverton

PENNSYLVANIA

Melrose Park Hamahreum
7320 Old York Road, Elkins Park
Yu's Oriental Food
1925 Cheltenham Avenue, Elkins Park

New Sambok Oriental Market
1737 Penn Avenue, Pittsburgh

SOUTH CAROLINA

Hyundai Oriental Market
1807 Decker Boulevar #1, Columbia

TENNESSEE

Asiana Market
5940 Mt. Moriah, Memphis
Hyundai Grocery Market
Memphis 1102 Charlotte Avenue, Nashville

Park & Shop Oriental Market
3664 Summer Avenue, Memphis
Tennessee Oriental Market

TEXAS

New Oriental Market
6929 Airport Boulevard #121, Austin
Shin Chon Oriental Market
11422 Harry Hines #210, Dallas
Sam Mi Oriental Market
6674 Arlington Boulevard, Falls Church
Chong-Song Meat Market
10082 Long Point Road, Houston
Dong Yang Market
8626 Long Point Road, Houston
Han-Yang Sik-Pum

Man-Na Sik-Pum
10001 Long Point Road, Houston

Sam-Bo Market
7655 Demoss Drive, Houston
Seoul Market
1809 Gessner Street, Houston

Irving Seoul Food
930 North Beltline Road #132, Irving
Poong Nyun Market
839 North Beltline Road, Irving
Richmond Oriental

Rx: Coconuts! (THE PERFECT HEALTH NUT)

5700 Telephone Road, Houston
Jin-Mee Sik-Pum
9501-Y Long Point Road, Houston

431 East Belt Blvd., Richmond
Lot-Tae Supper Market
7501 Harwin Drive, #101-E, Houston

UTAH

Oriental Food Market
667 South 700 East Street, Salt Lake City

VIRGINIA

Grand Mart
6255 Little River Turnpke, Alexandria
Annandale Korea House
4231 Markham Street, Annandale
Grand Mart-Centreville
5900 Centreville Crest Lane, Centreville
Dong-A Asian Market
9590 Lee Highway, Fairfax

Lotte Plaza
3250 Old Lee Highway, Fairfax

Han Ah Reum
8103 Lee Highway, Falls Church

Healthway Natural Foods
P.O. Box 129, Paeonian Springs
Kohyang Sikpum
6343-A Rolling Road, Springfield

Oriental Seoul Market
13662 Jefferson Davis Highway, Springfield

WASHINGTON STATE

Arirang Oriental Market
7940 Martin Way, Olympia

WISCONSIN

Tropical Traditions, Inc.
PMB #219-823 S. Main Street
West Bend

239

VERMÉN M. VERALLO-ROWELL, M.D.

Retailers and Suppliers of Specialized Coconut Products

MONOGLYCERIDES

MONOLAURIN—Lauricidin®

Contact: Dr. Jon Kabara
www.lauricdin.com

ENZYMATIC VIRGIN COCONUT OIL MONOGLYCERIDES

Contact: Dr. Teresita Espino
University of the Philippines, Los Banos, Philippines
TME@laguna.net
www.uplb.edu.ph/admin/OVCRE/Biotech

MONOLAURIN®/COCONUT MONOGLYCERIDES®

Hand And Foot Monolaurin Gel
Skin Sciences Research Center
Contact: Teresita G. Guttierres
    Unit 2012, Medical Plaza Makati,
    Amorsolo corner De la Rosa Streets, Legaspi Village,
    Makati, Philippines
    Tel + 63 2 811 2449
    Fax + 63 2 811 2465

LAURICIDIN®/COCONUT MONOGLYCERIDES® In

Hypoallergenic, Preservative-Free Cosmetics/ Skin Care Products
Id Sweat Acne and Overall Antibacterial Monolaurin® Gel
VMV Hypoallergenics

Contact: Laura Verallo de Bertotto
28/F Corporate Center
Sedeño cor. Valero Street
Makati City Philippines
Tel + 63 2 889 6137/39
Fax + 63 2 889 6140
www.vmvhypoallergenics.com

Suppliers: Where to Have Your Own Coconut Products Developed or Manufactured in the Philippines

Celebes Group of Companies, USDA Organic Certified
Farm Site: Butuan, Mindanao
Contact: Rory Ong Yiu
  No. 60, Lapu-Lapu, San Antonio Village, Apas
  Cebu City 6000 Philippines
  Tel + 63 85 3427777
  Fax + 63 85 3428888
  Email: *rory@celebes.com*

Elixirs Pacific International Corp
Farm Site: Tanauan, Batangas
Contact: Mr. Gil V. Obias
  No. 3 T.M. Kalaw St. Tierra Pura, Quezon City 1101
  Tel + 63 02 032 0587
  Email: *obiasgil@elixirspacific.com*

Fiesta Brands, Inc.
Farm Site: Northern Mindanao
Contact: Ms. Imelda Z. Gorospe
  1052 EDSA Magallanes Village Makati City, Philippines
  Tel + 63 2 63 2 851 0730 to 40
  Email: *dcnmarketing@fiesta-brands.com*

Manila Herbal & Essential Oils Co., Inc.
Sirawan Food Corporation (for SunGee Coconut Flour)
Farm Site: Km. 9 Sasa, Davao City
Contact: Ms. Jing C. Reyes
2281 P.Tamo Ext., Makati, Philippines
+ 63 2 888 3944/45
+ 63 2 816 7185

Nature's Blessings, Inc.
Farm Site: San Narciso, Quezon or 26103 Frampton Ave Harbor City, Ca, 90710
Contact Person: Cleve Figueroa Telefax (562) 869-1611 Tel (310) 257-0107
Suite C Penthouse, Padilla Building
Emerald Ave., Pasig City, Metro Manila
Tel: 00 63 2 635 5030 Fax; 00 63 631 9652
www.naturesblessings.com.ph

Peter Paul Philippines Corp.
Farm Site: Laguna, Bicol, Quezon suppliers
Contact: Mr. Jesulito P. Cornejo
3rd Floor, Unit D
The JMT Corporate Condominium
ADB Avenue, Ortigas Center 1657 Pasig City, Philippines
Tel + 63 2 633 5978 to 79
Email: *jesulito.cornejo@peterpaul.com.ph*

San Benito-The Farm
Farm Site: Tipacan, Lipa City, Batangas
Contact: Eckard and Perla Rempe, San Benito, Batangas
+ 63 2 696 3175
Email: *info@thefarm.com*

Tropicana Food Products, Inc.
Contact: Mr. Sing Tiu/ Ms. Jeannette Tiu
Km. 84 Maharlika Hi-way, San Pablo City, Laguna
Tel + 63 49 562 0089

Fax + 63 49 562 0088
Mobile 0920.921.3877
Email: *cocoking@msc.net.ph*

Malasimbo Extra Virgin Coconut Oil
Farm Site: Puerto Galera, Oriental Mindoro
Plant Site; Brgy. San Isidro, Balatero, Puerto Galera Oriental Mindoro
Contact: Mrs. Ara d'Aboville
Villa Malasimbo, Brgy. San Isidro, Balatero
Puerto Galera, Oriental Mindoro
Tel + 632 843 6211
Fax + 632 815 0176

Don Felipe's Farms, USDA Organic Certified, Organic Virgin Coconut Oil, Organic Premium Virgin Coconut Oil
Farm Site: Sto. Nino, Capoocan, Leyte
Contact: Mr. Glendon Rowell
28th Floor, Corporate Center, Sedeño Street, Salcedo Village
Makati City, Philippines
Tel +632 889 6137 or 889 6138 or 889 6139
Fax +633 889 6140
glendonrowell@hotmail.com

References

At the start of this book, I promised you a simple journey, that with the coconut and its products, you would end up with a change in your lifestyle towards enhancing your diet, to make you healthier, and better-looking. I hope I was able to convince you within the ten chapters of the book. If you are curious, and want to learn more, these are the articles or books to read.

(1) Kaunitz H, Slanetch CA, Johnson RE, Babayan VK, Barsky G. Relation of saturated, medium—and long-chain triglycerides to growth, appetite, thirst, and weight maintenance requirements. J Nutr 1958; 64: 513-524.

(2) Kaunitz H, Johnson RE, Cotton RH. Comparison of medium chain triglycerides and other fats in a reducing diet. Proc. Tenth Int'l Congress on Nutrition, Kyoto 1975; 63-64

(3) Kaunitz H. Long-term study on primates of milk substitutes containing coconut oil, soybean oil, and MCT. Proc. 9th Int'l Congress on Nutrition, Basel 1975; 4: 199-224

(4) Baba N, Bracco EF, Seylar J, Hashim SA. Enhanced thermogenesis and diminished deposition of fat in response to overfeeding with diets containing medium-chain triglycerides. J Am Soc Clin Nutr 1981; 14:624

(5) Bach AC, Babayan VK. Medium-chain triglycerides: an update. Am J Clin Nutr 1982; 36:950-962

(6) Babayan VK. Medium-chain triglycerides and structured lipids. Lipids 1987; 22: 417-20

(7) Enig MG. Know Your Fats: The Complete Primer For Understanding Nutrition Of Fats, Oils And Cholesterol. Third Edition. Bethesda Press: Sept 2002

(8) St-Onge MP, Jones PJH. Recent advances in nutritional sciences. Physiological effects of medium-chain triglycerides: potential agents in the prevention of obesity. The American Society for Nutritional Sciences J Nutr 2002; 132: 329-32

(9) Bray GA, Lee M, Bray TL. Weight gain of rats fed medium-chain triglycerides is less than rats fed long-chain triglycerides. Int J Obes 1980; 4: 27-32.

(10) Geliebter A, Torbay N, Bracco E, Hashim SA, Van Itallie TB. Overfeeding with medium-chain triglyceride diet results in diminished deposition of fat. Am J Clin Nutr 1983; 37: 1-4

(11) Mabayo RT, Furuse M, Murai A, Okumura JI. Interactions between medium-chain and long-chain triacylglycerols in lipid and energy metabolism in growing chicks. Lipids 1994; 29: 139-44

(12) Rothwell NJ, Stock MJ. Stimulation of thermogenesis and brown fat activity in rats fed medium chain triglyceride. Metabolism 1987; 36: 128-30

(13) Seaton TB, Welle SL, Warenko MK, Campbell RG. Thermic effect of medium-chain and long-chain triglycerides in man. Am J Clin Nutr 1986; 44: 630-34

(14) Hill, J. O., Peters, J. C., Yang, D., Sharp, T., Kaler, M., Abumrad, N. N. & Greene, H. L. (1989) Thermogenesis in humans during overfeeding with medium-chain triglycerides. Metabolism 38: 641-648. 8.

(15) Scalfi L, Coltorti A, Contaldo F. Postprandial thermogenesis in lean and obese subjects after meals supplemented with medium-chain and long-chain triglycerides. Am J Clin Nutr 1991; 53: 1130-33

(16) Lasekan JB, Rivera J, Hirvonen MD, Keesey RE, Ney DM.

Energy expenditure in rats maintained with intravenous or intragastric infusion of total parenteral nutrition solutions containing medium—or long-chain triglyceride emulsions. J Nutr 1992; 122: 1483-92

(17) Dulloo, A. G., Fathi, M., Mensi, N. & Girardier, L. (1996) Twenty-four hour energy expenditure and urinary catecholamines of humans consuming low to-moderate amounts of medium-chain triglycerides: a dose-response study in human respiratory chamber. Eur. J. Clin. Nutr. 50: 152-158.

(18) St-Onge MP, Bourque C, Papamandjaris AA, Jones PJH. Consumption of medium chain triglycerides versus long chain triglycerides over 4 weeks increases energy expenditure and fat oxidation in obese women. Ann Nutr Metab 2001; 45 (1): 89

(19) Maggio CA, Koopmans HS. Food intake after intragastric meals of short-, medium-, or long-chain triglycerides. Physiol Behav 1982; 28:921-261

(20) Denbow DM, Van Krey HP, Lacy MP, Watkins BA. The effect of triacylglycerol chain length on food intake in domestic fowl. Physiol Behav 1992; 51: 1147-50

(21) Furuse M, Choi YH, Mabayo RT, Okumura JI. Feeding behavior in rats fed diets containing medium chain triglyceride. Physiol Behav 1992; 52: 815-17

(22) Stubbs RJ, Harbron CG. Covert manipulation of the ratio of medium—to long-chain triglycerides in isoenergetically dense diets: effect on food intake in ad libitum feeding men. Int J Obes 1996; 20: 435-44

(23) Van Wymelbeke V, Himaya A, Louis-Sylvestre J, Fantino M. Influence of medium-chain and long-chain triacylglycerols on the control of food intake in men. Am J Clin Nutr 1998; 68: 226-34

(24) http://www.monticello.org/gardens/vegetable/science_gardener.html

(25) Whitney EN, Rolfes SR. Understanding Nutrition. Thomson/Wadsworth, 9th Edition 2002

(26) Mensink RPM, Katan MB. Effect of dietary trans fatty acids on high-density and low density lipoprotein cholesterol levels in healthy subjects. N Engl J Med 1990; 323: 439-45

(27) Troisi R, Wllett WC, Wess ST.Trans-fatty acid intake in relation to serum lipid concentrations in adult men. Am J Clin Nutr 1992; 56(6): 1019-24

(28) Willet WC, Stampfer MJ, Manson JE, Colditz GA, Speizer FE, Rosner BA, Sampson LA, Hennekens CH. Intake of trans fatty acids and risk of coronary heart disease among women. Lancet 1993; 351: 581-85

(29) Judd JT, Clevidence BA, Muesing RA, Wittes J, Sunki ME, Podczasy JJ. Dietary trans fatty acids: effects on plasma lipids and lipoproteins of healthy men and women. Am J Clin Nutr 1994; 59: 861-68

(30) Expert panel on trans fatty acids and coronary heart disease. Trans fatty acids and coronary heart disease risk. Am J Clin Nutr 1995; 62: 6555-708S

(31) Clevidence VA, et al. Plasma lipoprotein (a) levels in men and women consuming diets enriched in saturated, cis-, or trans-monousaturated fatty acids. Arterioscler Thromb Vasc Biol 1997; 17: 1657-61

(32) Nelson GJ. Dietary fat, trans fatty acids, and risk of coronary heart disease. Nutr Rev 1998; 56: 250-52

(33) Sebedio JL, Christie WW eds. In: Trans Fatty Acids in Human Nutrition, Volume 9 in the Oil Press Lipid Library. The Oily Press: Dundee, Scotland 1998

(34) Federal Register Final Rule: Trans-fatty acids in Nutrition Labeling, Nutrient Content, Claims and Health Claims, Dept. of Health, Food and Drug Administration July 11, 2003.

(35) Smith LM, Dunkley WL, Franke A, Dairiki T. Measurement of trans and other isomeric unsaturated fatty acids in butter and margarine. J Am Oil Chem Soc 1978; 55: 257-61

(36) Enig MG, Pallansch LA, Sampugna J, Keeney M. Fatty acid composition of fat of selected food items with emphasis on trans components. J Am Oil Chem Soc 1983; 60 (10): 1788-95

(37) Wahle KW, James WPT. Isomeric fatty acids and human health. Eur J Clin Nutr 1983; 47: 828-39

(38) Gurr MI. Trans-fatty acids: metabolic and nutritional significance. Bull Int Dairy Fed 1990; 166: 5-18

(39) Bakery foods are the major dietary source of trans-fatty acids among pregnant women with diets providing 30 percent energy from fat. J Am Dietetic Ass 2002; 102 (1): 46-51

(40) Francois CA, Connor SL, Wander RC, Connor WE. Acute effects of dietary fatty acids on the fatty acids of human milk. Am J Clin Nutr 1998; 67: 301-08

(41) Koletzko B. Supply, metabolism and biological effects of trans-isomeric fatty acids in infants. Nahrung 1991; 35: 229-83

(42) Koletzko B. Potential adverse effects of trans fatty acids in infants and children. Eur J Med Res 1995; 1: 123-5

(43) Enig MG. Research review: trans fatty acids-an update. Nutr Quarterly 1993; 17 (4): 79-95

(44) Enig MG, Atal S, Keeney M, Sampugna J. "Isomeric trans fatty acids in the U.S. diet" J Am Coll Nutr 1990; 9: 471-86

(45) Ratnayake WMN, Hollywood R, O'Grady E, Pelletier G. Fatty acids in some common food items in Canada. J Am Coll Nutr 1993; 12: 651-60

(46) Santayana, George. Accessed: Apr. 2005 http://www.britannica.com

(47) Enig, Mary G, PhD, Trans Fatty Acids in the Food Supply: A Comprehensive Report Covering 60 Years of Research, 2nd Edition, Enig Associates, Inc, Silver Spring, MD, 1995, 148-154; Enig, Mary G, PhD, et al, J Am Coll Nutr, 1990, 9:471-86

(48) Shelton EM Jr. The coconut industry and the excise tax. The Am Cham of Commerce (Phils). Vol. XIV No. 2. Dec 1934

(49) Speech of Pres. Manuel L. Quezon on the Excise Tax on Coconut Oil, National Language and Social Justice, February 19,1938

(50) Enig MG. The Tragic Legacy of CSPI/ Wise Traditions in Food, Farming and the Healing Arts, the quarterly magazine of the Weston A. Price Foundation, Fall 2003

(51) Enig M. Health and Nutritional Benefits from Coconut Oil: An Important Functional Food for the 21st Century Presented at the AVOC Lauric Oils Symposium, Ho Chi Min City, Vietnam, 25 April 1996
(52) Cardiovascular epidemiology: historical perspectives and assessing risk of CVD. www.heart.uci.edu accessed April 8, 2005
(53) Keys A. Atherosclerosois: A problem in newer public helath. Journal of Mount Sinai Hospital 20, 118-139, 1953
(54) Kaunitz H, Dayrit CS. Coconut oil consumption and coronary heart disease. Coconuts Today special issue October 22, 2001
(55) Kaunitz H. Nutritional properties of coconut oil. J Am Oil Chem Soc 1970; 47: 462A-464A. 482A-485A
(56) Mendis GS, Wissler RW, Brindenstein RT, Podielski FJ. The effects of replacing coconut oil with corn oil on human serum lipid profiles and platelet derived factors active in atherogenesis. Nutr Rep Int 1989 Oct; 40: 4
(57) Prior IAM. Cardiovascular epidemiology in New Zealand and the Pacific. NZ Med J 1974; 80: 245-52
(58) Stanhope JM, Sampson V, Prior IAM. The Tokelau Island migrant study: serum lipid concentrations in two environments. J Chron Dis 1981; 34: 45
(59) Prior IA, Davidson F, Salmond CE, Czochanska Z. Cholesterol, coconuts, and diet on Polynesian atolls-a natural experiment: the Pukapuka and Tokelau island studies. Am J Clin Nutr 1981; 34: 1552-61
(60) Florentino RF, Aguinaldo AR. Diet and cardiovascular disease in the Philippines. Phil J Coconut Studies 1987; 12: 56-70
(61) Ng TKW, Hassan K, Lim JB, Lye MS, Ishak R. Non-hypercholesterolemic effects of a palm oil diet in Malaysian volunteers. Am J Clin Nutr 1991; 53: 1015S-1020S
(62) Kurup PA, Rajmohan TH. Consumption of coconut oil and coconut kernel and the incidence of atherosclerosis. Human Nutrition Proceedings. Symposium on Coconut and Coconut Oil in Human Nutrition 27 March 1994; Coconut Development Board, Kochi, India 1995; 35-59

(63) Eraly MG. IV. Coconut oil and heart attack. Coconut and Coconut Oil in Human Nutrition, Proceedings. Symposium on Coconut and Coconut Oil in Human Nutrition. 27 March 1994. Coconut Development Board, Kochi, India, 1995, pp 63-64.

(64) Kumar PD. The role of coconut and coconut oil in coronary heart disease in Kerala, South India. Trop Doct 1997 Oct; 27 (4): 215-17

(65) Sircar S, Kansra U. Choice of cooking oils—myths and realities. J Indian Med Assoc 1998 Oct; 96 (10): 3040-7

(66) Nevin KG, Rajmohan T. Lipid lowering effect of virgin coconut oil. In: National Seminar on Coconut and Coconut products in health and disease Sept.19-20 2003; 126-28. 10 Kanakakunnu Palace, Thiruvananthapuram, India.

(67) Rao, Ananth N, Vasudevan DM. Nutritional relevance of coconut oil-myth versus reality, Sept.19-20 2003; 77-80 in National Seminar on Coconut Products in Health and disease, Kanakakunnu Palace, Thiruvananthapuram, India

(68) Kintanar QL, Castro JS. Is coconut oil hypercholesterolemic and atherogenic? A focused review of the literature. Trans Nat Acad Sci Tech (Phils) 1988 10: 371-414

(69) Malmros H, Wigand G. The effects on serum cholesterol of diets containing different fats. Lancet 1957; 2: 1-8

(70) Hashim SA, Clancy RE, Hegsted DM, Stare FJ. Effect of mixed fat formula feeding on serum cholesterol level in man. Am J Clin Nutr 1959; 7: 30-34

(71) Halden VW, Lieb H. Influence of biologically improved coconut oil products on the blood cholesterol levels of human volunteers. Nutr Dieta 1961; 3: 75-88

(72) Reiser R. Saturated fat in the diet and serum cholesterol concentration: a critical examination of the literature. Am J Clin Nutr 1973; 26: 524-55

(73) Blackburn GL, Kater G, Mascioli EA, Kowalchuk M, Babayan VK, Bistrian BR. A reevaluation of Coconut oil's effect on serum cholesterol and atherogenesis. Coconuts Today. Special Issue, for the 13th Asian Pacific Congress of Cardiology, Manila Feb 11-16, 1990

(74) de Roos NM, Schouten EG, Martijn KB. Consumption of a Solid Fat Rich in Lauric Acid Results in a More Favorable Serum Lipid Profile in Healthy Men and Women than Consumption of a Solid Fat Rich in trans-Fatty Acids. J Nutr 2001;131:242-245

(75) Ronald P Mensink, Peter L Zock, Arnold DM Kester and Martijn B Katan. Effects of dietary fatty acids and carbohydrates on the ratio of serum total to HDL cholesterol and on serum lipids and apolipoproteins: a meta-analysis of 60 controlled trials Amer J Clin Nutr: 77; 5, 1146-1155, May 2003

(76) German JB and Dillard CJ. Commentary: Saturated fats: what dietary intake? American Journal of Clinical Nutrition, Vol. 80, No. 3, 550-559, September 2004.

(77) Felton CV, Crook D, Davies MJ, Oliver MF. Dietary polyunsaturated fatty acids and composition of human aortic plaques. Lancet, 344:1195-1196;1994.

(78) Dietary guidelines for healthy American adults. American Heart Association 1986

(79) Atherosclerosis Study Group. Primary prevention of the atherosclerotic diseases. Circulation 1970; 42 (1): 1-95

(80) USDA Food Pyramid: www.usda.gov/cnpp/FENR/V11N4/fenrv11n4p42.PDF

(81) Incidence hypertension, heart disease, diabetes http://www.surgeongeneral.gov/topics/obesityI accessed April 9, 2005

(82) James PT, Leach R, Kalamara E, Shayeghi M. The worldwide obesity epidemic. Section I: Obesity, the Major Health Issue of the 21st Century Obesity Research 2001; 9: S228-S233.

(83) Enig MG. Diet, serum cholesterol and coronary heart disease, in Mann GV (ed): Coronary Heart Disease: The Dietary Sense and Nonsense. Janus Publishing: London 1993; 36-60

(84) Raanskov U. The cholesterol Myths. Exposing the fallacy that saturated fat and cholesterol cause heart disease. Sept 2000; New Trends Publishing

(85) Dayrit CS. On atherosclerosis and diseases of degeneration. Trans Natl Acad Sci and Tech Phil 2001; 23: 163-77

(86) McGregor L. Effect of feeding with hydrogenated coconut oil on platelet function in rats. Abstract Proc Nutr Soc 1974; 33: 1A-2A

(87) Tannenbaum A. The dependence of the genesis of induced skin tumors on the fat content of the diet during different stages of carcinogenesis. 1944; Cancer Res. 4; 633-687

(88) Lavik PS, Baumann CA. Further studies on tumor-promoting action of fats. 1979; Cancer Res 749-756

(89) Tannenbaum A. 1947. Effect of varying caloric intake upon tumor incidence and tumor growth. Ann. N.Y. Acad sci 49.5

(90) Carroll KK, Khor HT. Effects of dietary fat and dose level of 7, 12-dimethylbenz(alpha)-anthracene on mammary tumor incidence in rats. m1975; 30: 226

(91) Reddy BS, Weisburger JH, Wynder EL. Effect of dietary fat level and dimethylhydrazine on fecal acid and sterol excretion and colon carcinogenesis in rats. J Natl Cancer Inst 1974; 52: 507-11

(92) Chan PC, Dae TL. Enhancement of mammary carcinogenesis by a high-fat diet in Fischer, Long Evans and Sprague-Dawley rats. Cancer Res 1981 Jan; 41 (1): 164-7

(93) Bull AW, Soullier BK, Wilson PS, Hayden MT, Nigro ND. Promotion of azoxymethane-induced intestinal cancer by high-fat diets in rats. Cancer Res 1979 Dec; 39: 1363

(94) Pariza MW. Dietary fat, calorie restriction, ad libitum feeding and cancer risk. Nutr Rev 1987; 45: 1-7

(95) Baumann CA, Rusch HP. Effect of diet on tumors induced by ultraviolet light. Am J Cancer. 1939;35:213-221.

(96) Black HS.Influence of dietary factors on actinically-induced skin cancer. Mutation Research/ Fundamental and Molecular Mechanisms of Mutagenesis 1998 Nov; 422(1):185-190

(97) Black HS, Herd JA, Goldberg LH, Wolf JE, et al. Effect of a low fat diet on the incidence of actinic keratosis. New Engl J Med 1992; 1272-75

(98) Lim-Sylianco CY. Anticarcinogenic effect of coconut oil.The Philippine Journal of Coconut Studies Proceedings of the

Symposium on "Medical and Nutritional Aspects of Coconut Oil"1987 December; XII No.2: 89-102
(99) Opie EL 1944. the influence of diet in the production of tumors of the liver by butter yellow. J Exper, ed 80-:219-230
(100) Klline BE et al 1946. The carcinogenicity of dimethylaminoazzobenzene in diets containing fatty acids of coconut oil or of corn oil
(101) Miller JA et al 1944. The effect of certain lipids on the carcinogenicity of p-dimethylaminioazobenzene. Cancer Res. 4:756=761
(102) Broitman SA et al. Polyunsaturated fat, cholesterol, and large bowel carcinogenesis. 1977;Cancer 40:245
(103) Reddy BS, Maorua Y. Tumor promotion of dietary fat in azoxymethane-induced colon carcinogenesis in female F 344 rats. 1984; J Natl Cancer Inst 72: 745
(104) Gammal EB, Caroll KK, Plunkett ER. Effect of dietary fat on mammary carcinogenesis by 7,12-dimethylbenz(a)anthracene. 1967; Cancer Res. 27: 1737-1742
(105) Caroll KK, Khor MT. Effects of dietary fat and dose level of 7,12-dimethylbenz(a)anthracene on mammary tumor incidence in rats. 1970; Cancer Res. 30:2260
(106) Chan PC, Dae TL. 1981. Enhancement of mammary carcinogenesis by a high-fat diet in Fischer, Long, Evans and Sprague-Dawley rats. Cancer Res 41:164-167
(107) Cohen LA, Thompson DO, Maeura Y, Choi K, Blank M, Rose DP. Dietary fat and mammary cancer. I. Promoting effects of different dietary fats on N nitrosomethylurea-induced rat mammary tumorigenesis. J Nat Cancer Inst 1986; 77: 33
(108) Roebuck BD, et al. Promotion of unsaturated fat of azaserine-induced pancreatic carcinogenesis in the rat.1981; Cancer Res. 41: 3961
(109) Carroll KK. Diet and carcinogenesis: historical perspectives. Adv Exp Med Biol 1986;206:45-53.
(110) Hillyard LA, Abraham S. Effect of dietary polyunsaturated fatty acids on the growth of mammary adenocarcinoma in mice and rats. Cancer Res 1979; 39: 4430-3

(111) Karmali, R A, Marsh J, Fuchs C. 1984. Effect of omega-3 fatty acids on growth of a rat mammary tumor. J Natl Cancer Inst., 73: 457-461.

(112) Karmali R A 1987. Eicosanoids in neoplasia. Prev Med. 16:493-502

(113) Mimoura T et al. 1988. Effect of dietary eicosapentaenoic acid (EPA)on azoxymethane-induced colon carcinogenesis in rats. Cancer Res. 48: 4790-4794

(114) Lindner MA. A fish oil diet inhibits colon cancer in mice. Nutr Cancer 1991;15:1-11. [Medline]

(115) Rose DP 1997. Dietary fatty acids on breast and prostate cancer: evidence from in vitro experiments and animal studies. Am J Clin Nutr 66 (suppl): 1413S-1422S

(116) Fay MP, Freedman LS, Clifford CK, and Midthune DN. Effect of Different types and amounts of fat on the development of mammary tumors in rodents: a review. Cancer Research1997; 3979-3988

(117) Cave WT Jr. Omega-3 polyunsaturated fatty acids in rodent models of breast cancer. Breast Cancer Res Treat. 1997 Nov-Dec;46(2-3):239-46.

(118) Takahashi M. et al1997: Suppression of rat colon carcinogenesis by docosahexanoic acid (DHA). Proc Am Assoc cancer Res 38: 109

(119) Hubbard NE, Lim D, Erickson KL 1998. Alteration of murine mammary tumorigenesis by dietary enrichment with omega-3 fatty acids in fish oil Cancer Lett 124:1-7

(120) Ip C 1997 Review of the effects of trans fatty acids, oleic acid, n-3 polyunsaturated fatty acids, and conjugated linoleic acid on mammary carcinogenesis in animals. Am J Clin Nutr 66 (suppl) 1523 S01529 S

(121) Howe GR et al. 1990. Dietary factors and risk of breast cancer: combined analysis of 12 case control studies. J Natl Cancer Inst 82: 561-569

(122) Anti M, Armelao F, Marra G, et al. Effects of different doses of fish oil on rectal cell proliferation in patients with sporadic colonic adenomas. Gastroenterology 1994;107:1709-18

(123) Favero A, Parpinel M, Franceschi S. Diet and risk of breast cancer: major findings from an Italian case-control study. Biomed Pharmacother 1998;52:109-15.

(124) Greenwald P 1999. Role of dietary fat in the causation of breast cancer: point. Cancer Epidemiol Biomarkers Prev., 8:3-7

(125) Norrish AE, Skeaff CM, Arribas GL, Sharpe SJ, Jackson RT. Prostate cancer risk and consumption of fish oils: a dietary biomarker-based case-control study. Br J Cancer 1999;81:1238-42

(126) Franceschi S, Favero A, La Vecchia C, et al. Influence of food groups and food diversity on breast cancer risk in Italy. Int J Cancer 1995;63:785-9.

(127) Rose DP, Connolly JM. Omega-3 fatty acids as cancer chemopreventive agents. Pharmacol Ther 1999; 83: 217-44

(128) Terry P, Lichtenstein P, Feychting M, Ahlbom A, Wolk A. Fatty fish consumption and risk of prostate cancer. Lancet 2001;357:1764-6

(129) Terry P, Wolk A, Vainio H, Weiderpass E. Fatty fish consumption lowers the risk of endometrial cancer: a nationwide case-control study in Sweden. Cancer Epidemiol Biomarkers Prev 2002;11:143-5.

(130) Augustsson K, Michaud DS, Rimm EB, et al. A prospective study of intake of fish and marine fatty acids and prostate cancer. Cancer Epidemiol Biomarkers Prev 2003;12:64-7.

(131) M Gago-Dominguez1, J-M Yuan1, C-L Sun1, H-P Lee2 and M C Yu1. Opposing effects of dietary n-3 and n-6 fatty acids on mammary carcinogenesis: The Singapore Chinese Health Study British Journal of Cancer (2003) 89, 1686-1692doi:10.1038/sj.bjc.6601340

(132) Hunter DJ et al1996. Cohort studies of fat intake and the risk of breast cancer-a pooled analysis. N Engl J Med 334:356-361.

(133) Holmes MD, et al. 1999 Association of dietary intake of fat and fatty acids with risk of breast cancer. J Am Med Assoc 281: 914-920

(134) Zock PL, Katan MB. Linoleic acid intake and cancer risk. A review and meta-analysis. Am J Clin Nutr 1998;68:142-53. 13.

(135) Erickson KL 1998. is there a relation between dietary linoleic acid and cancer of the breast, colon, or prostate? Am J Clin Nutr. 68:5-7

(136) Chajes V, Hulten K, Van Kappel AL, et al. Fatty-acid composition in serum phospholipids and risk of breast cancer: an incident case-control study in Sweden. Int J Cancer 1999;83:585-90. [Medline]

(137) Holmes MD, Hunter DJ, Colditz GA, et al. Association of dietary intake of fat and fatty acids with risk of breast cancer. JAMA 1999;281:914-20.

(138) Severson RK, Nomura AM, Grove JS, Stemmermann GN. A prospective study of demographics, diet, and prostate cancer among men of Japanese ancestry in Hawaii. Cancer Res 1989;49:1857-60.

(139) Vatten LJ, Solvoll K, Loken EB. Frequency of meat and fish intake and risk of breast cancer in a prospective study of 14,500 Norwegian women. Int J Cancer 1990;46:12-5

(140) Hursting SD, Thornquist M, Henderson MM. Types of dietary fat and the incidence of cancer at five sites. Prev Med 1990;19:242-53

(141) Galli C, Butrum R. Dietary three fatty acids and cancer. An Overview. World Rev. Nutr Diet 1991; 66: 446-61

(142) Enstrom JE. Reassessment of the role of dietary fat in cancer etiology. Cancer Res 1998; 41(9): 373

(143) Kolonel LN, Nomura AM, Cooney RV. Dietary fat and prostate cancer: current status. J Natl Cancer Inst 1999; 91: 414-28

(144) Willett WC. Specific fatty acids and risks of breast and prostate cancer: dietary intake. Am J Clin Nutr 1997; 66: 1557S-63S

(145) Willett WC. Dietary fat intake and cancer risk: a controversial and instructive story. Semin Cancer Biol 1998; 8: 245-53

(146) Bartsch H, Nair J, Owen RW. Dietary polyunsaturated fatty acids and cancers of the breast and colorectum: emerging evidence for their role as risk modifiers. Carcinogenesis. 1999 Dec; 20 (12): 2209-18

(147) Terry PD, Rohan TE, Wolk A. Intakes of fish and marine fatty acids and the risks of cancers of the breast and prostate and of

other hormone-related cancers: a review of the epidemiologic evidence. Am J Clin Nutr 2000 Mar; 77 (3): 532-543

(148) Larsson SC, Kumlin M, Ingelman-Sundberg M, Wolk A. Dietary long-chain n-3 fatty acids for the prevention of cancer: a review of potential mechanisms. American Journal of Clinical Nutrition 2004 June; 79 (6): 935-45

(149) Hilakivi-Clarke L et al 1996. Breast cancer risk in rats fed a diet high in n-6 polyunsaturated fatty acids during pregnancy. J Natl Cancer Inst; 88: 1821-1827

(150) Hilakivi-Clarke L et al 1997. A maternal diet high in n-6 polyunsaturated fats alters mammary gland development, puberty onset, and breast cancer risk among female rat offspring. Proc Natl Acad Sci USA 94: 9372-9377

(151) Walker BE and Kurth LE. Multigenerational effects of dietary fat carcinogenesis in mice. Cancer Res. 1997 Oct 1;57(19):4162-3.

(152) Kaizer L, Boyd NF, Kriukov V, Tritchler D. Fish consumption and breast cancer risk: an ecological study. Nutr Cancer 1989;12:61-8.

(153) Caygill CP, Charlett A, Hill MJ. Fat, fish, fish oil and cancer. Br J Cancer 1996;74:159-64

(154) Sasaki S. Horacsek M, Kesteloot H. An ecological study of the relationship between dietary fat intake and breast cancer mortality. Prev Med 1993; 22: 1887-202

(155) Ishimoto H, Nakamura H, Miyoshi T.Tokushima. Epidemiological study on relationship between breast cancer mortality and dietary factors. J Exp Med. 1994 Dec;41(3-4):103-14.

(156) Kiyonori Kuriki et al Plasma Concentrations of (n-3) Highly unsaturated fatty acids are good biomarkers of relative dietary fatty acid intakes: A cross-sectional study 2003 American Society for Nutritional Sciences J. Nutr. 133:3643-3650,

(157) Kobayashi M, Sasaki S, Hamada GS, Tsugane S.Serum n-3 fatty acids, fish consumption and cancer mortality in six Japanese populations in Japan and Brazil. Jpn J Cancer Res. 1999 Sep;90(9):914-21.

(158) Lands WE, Hamazaki T, Yamazaki K, et al. Changing dietary patterns. Am J Clin Nutr 1990; 51: 991-3
(159) Eiliv Lund and Kaare H. Bønaa. Reduced breast cancer mortality among fishermen's wives in Norway. May 1993 Cancer Causes and Control (Historical Archive) Issue: Volume 4, Number 3 283-287
(160) Terry P, Wolk A, Vainio H, Weiderpass E. Fatty Fish Consumption Lowers the Risk of Endometrial Cancer. A Nationwide Case-Control Study in Sweden. Cancer Epidemiology Biomarkers & Prevention; 11: 143-145, January 2002
(161) Lanier AP, Kelly JJ, Smith B, et al. Alaska Native cancer update: incidence rates 1989-1993. Cancer Epidemiol Biomarkers Prev 1996;5:749-51.[Abstract]
(162) Nielsen NH, Hansen JP. Breast cancer in Greenland—selected epidemiological, clinical, and histological features. J Cancer Res Clin Oncol 1980;98:287-99.[Medline]
(163) David M. Eisenberg et al. Trends in Alternative Medicine Use in the United States, from 1990 to 1997: Results of a Follow-up National Survey. JAMA 1998; 280 (18) Nov 11: 1569-1575
(164) Chajes V, Sattler W, Stranzl A, Kostner GM. Influence of n-3 fatty acids on the growth of human breast cancer cells in vitro: relationship to peroxides and Vitamin-E. Breast Cancer Res Treat 1995; 34: 199-212
(165) Lhullery C, Cognault S, Germain E, Jourddan ML, Bougnoux P. Suppression of the promoter effect of polyunsaturated fatty acids by the absence of dietary vitamin E in experimental mammary carcinoma. Cancer Lett 1997; 114: 233-34
(166) Lhuillery C, Bougnous P, Groscolas R, Durand G 1995. Time-course study of adipose tissue fatty acid composition during mammary tumor growth in rats with controlled fat intake. Nutrition and Cancer 24. 299-309
(167) Capone SL, Bagga D, Glaspy J A 1997. Relationship between omega-3 and omega-6 fatty acid ratios and breast cancer. Nutrition 13, 822-823
(168) Maillard V, Bougnoux P, Ferrari P, et al. n-3 and n-6 fatty acids in breast adipose tissue and relative risk of breast cancer in a

case-control study in Tours, France. Int J Cancer 2002;98:78-83.

(169) Eaton, S. B., Eaton, S. B., III, Sinclair, A. J., Cordain, L. & Mann, N. J. (1998) Dietary intake of long-chain polyunsaturated fatty acids during the Paleolithic. World Rev. Nutr. Diet. 83:12-23.

(170) Simopoulos A P. The Mediterranean Diets: What is so special about the diet of Greece? The scientific evidence 2001. J Nutr 131:3065S-3073S

(171) De Lorgeril M, Salen P. 2000 Modified Cretan Mediterranean diet in the prevention of coronary heart disease and cancer. Simpoulos AP, Visoli F eds. Mediterranean Diets 87:1-23 Karger Basel, Switzerland

(172) Solanas M—Effects of a high olive oil diet on the clinical behavior and histopathological features of rat DMBA-induced mammary tumors compared with a high corn oil diet. Int J Oncol—01-Oct-2002; 21(4): 745-53

(173) Menendez J, Meyers M, Brooks J. Jan. 10, 2005, Annals of Oncology

(174) Martin-Moreno JM, Willett, WC, et al. 1994. Dietary fat, olive oil intake and breast cancer risk. Int. J Cancer. 58:774-780

(175) Wolk A, et al. A Prospective Study of Association of Monounsaturated Fat and Other Types of Fat With Risk of Breast Cancer. Arch Intern Med. 1998; 158: 41-45

(176) LaVecchia C, Negri E, Franceschi S, Decarli A, Giacosa A, Lipworth I. 1995. Olive oil, other dietary fats and the risk of breast cancer (Italy) Cancer Causes Control 6:545-555

(177) Trichopoulou A, Katsouyanni K, et al. 1995;Consumption of olive oil and specific food groups in relation to breast cancer risk in Greece.

(178) Lipworth L, Martinez M E, Angell J, Hsieh C C, Trichopoulos, D1997. Olive oil and human cancer: an assessment of evidence. Preventive Medicine 26, 181-190

(179) Assmann G, de Backer G, et al. 1997. International consensus statement on olive oil and the Mediterranean diet; implications for health in Europe. Eur J Cancer Prev; 6: 418-421

(180) Simonsen NR, et al. 1998: Tissue stores of individual monounsaturated fatty acids and breast cancer. The EURAMIC study. Am J Clin Nutr 68: 134-

(181) Takashita M, Ueda H, Shirabe K, Higuci Y, Yoshida S. 1997. Lack of promotion of colon carcinogenesis by high-oleic safflower oil. Cancer 79: 1487-1493

(182) Hillyard LA, Abraham S.1979. Efect of dietary polyunsaturated fatty acids on the growth of mammary adenocarcinoma in mice and rats.

(183) Lim-Sylianco CY, Balboa J, Casareno R, Mallorca R, Serrame E, Wu LS. Antigenotoxic effects on bone marrow cells of two vegetable oils. Phil J Coconut Studies 1992 Dec; 18 (9): 6-10

(184) Lim-Sylianco CY, Mallorca R, Serrame E, Wu LS. A comparison of germ cell antigenotoxic activity of nondietary and dietary coconut oil and soybean oil. Phil J Coconut Studies 1992 Dec; 18 (9): 1-5

(185) Tappel AL. Chapter III in Pathology of Cell membranes. Trump BF, Arstilla A eds, Academic Press: New York 1975

(186) James MJ, Gibson RA, Cleland LG. Dietary polyunsaturated fatty acids and inflammatory mediator production. Am J Clin Nutr 2000 Jan; 71 (1): 343S-348s

(187) Dommels YE, Haring MM, Keestra NG, Alink GM, van Bloaderen PJ, van Ommen B. The role of cyclooxygenase in n-6 and n-3 polyunsaturated fatty acid mediated effects on cell proliferation, PGE (2) synthesis and cytotoxicity in human colorectal carcinomal cell lines. Carcinogenesis 2003; 24: 385-92

(188) Calder PC, Yaqoob P, Thies F, Wallace FA, Miles EA. Fatty acids and lymphocyte functions. Br J Nutr 2002;87 (suppl):S31-48

(189) Morrison JH, Svoboda R. The Book of Ayurveda: A Holistic Approach to Health and Longevity. Simon & Schuster Inc.: New York 1995, A Fireside Book.

(190) Medicinales de Filipinas 1892, Reprinted by Ayala Foundation Inc., 2000, Makati, Philippines

(191) Quisumbing E. Medicinal Plants of the Philippines. JMC Press, Inc. Manila, Philippines 1978

(192) Sircar S, Kansra U. Choice of cooking oils—myths and realities. J Indian Med Assoc 1998 Oct; 96 (10): 304-07

(193) Anzaldo FE, Kintanar QL, Recio PM, Velasco RU, dela Cruz F, Jacalne A. Coconut water as intravenous fluid. Phil J Pediatrics 1975 Aug; 24: 143-66.

(194) Intengan CL, Pesigan JS, Cawaling T, Zalamea IY, Dayrit CS. Structured lipids of coconut oil and corn oil vs. soybean oil in the rehabilitation of malnourished children—A Field Study. Philippine Journal of Internal Medicine 1992 July/August; 30

(195) Macalalag EV, Jr. et al. Buko water of immature coconut is a universal urinary stone solvent. Read at the Pacific Coconut Community Conference, Legend Hotel, Metro Manila, August 14-18, 1997.

(196) Sachs M, von Eichel J, Askali F. Coconut oil used in Indonesia for treatment of wounds and to preserve medicinal herbs. Chirurg 2002 Apr; 73 (4): 387-92

(197) Kabara JJ, Swieczkowski DM, Conley AJ, Truant JP: Fatty acids and derivatives as antimicrobial agents. Antimicrob Agents Chemother 1972; 23: 28

(198) Kabara JJ, Vrable R, Lie Ken Jie M. Antimicrobial lipids: natural and synthetic fatty acids and monoglycerides. Lipids 1977; 9: 753

(199) Kabara JJ. Fatty acids and derivatives as antimicrobial agents— A review, in The Pharmacological Effect of lipids. In: Kabara JJ ed. American Oil Chemists' Society, Champaign IL 1978; 1-14 Kabara JJ.

(200) Kabara JJ. Toxicological, bacteriocidal and fungicidal properties of fatty acids and some derivatives. JAOCS 1979; 56: 76

(201) Kato N. Combined effects on antibacterial activity of fatty acids and their esters against gram-negative bacteria. In: Kabara JJ ed. The pharmacological effects of lipids. St. Louis, Mo. Am Oil Chem Soc 1978; 15: 24

(202) Kabara JJ. Inhibition of staphylococcus aureus in The Pharmacological Effect of Lipids II (Kabara JJ ed.) Am Oil Chem Soc, Champaign IL 1985; 71-75

(203) Fletcher RD, Albers AC, Albertson JN, Kabara JJ. Effects of monoglycerides on mycoplasma pneumoniae growth. In: Kabara JJ ed. The Pharmacological Effect of Lipids II Am Oil Chem Soc, Champaign IL 1985; 59-63

(204) Kabara JJ. GRAS antimicrobial agents for cosmetic products. J Soc Cosmet Chem 1980; 31: 1-10

(205) Kabara JJ. Lauricidin: the non-ionic emulsifier with antimicrobial properties. Cosmet Sci Technol 1984; I: 305-322

(206) Kabara JJ. Chemistry and biology of monoglycerides. In: A Key to Cosmetic Ingredients. Marcel Dekker Inc.: New York 1991; 311-20.

(207) Kabara JJ. Food-grade chemicals for use in designing food preservatives. J Food Prot 1981; 44: 633-647.

(208) Kabara JJ. Inhibition of staphylococcus aureus in a model sausage system by monoglycerides. 1985; 71-103. In: Kabara JJ (ed). The pharmacological effect of lipids II The Am Oil Chem Soc, Champaign, Ill

(209) Hierholzer JC, Kabara JJ. In vitro effects of monolaurin compounds on enveloped RNA and DNA viruses. J Food Safety 1982; 4: 1-12.

(210) Boddie RL, Nickerson SC. Evaluation of postmilking teat germicides containing Lauricidin, saturated fatty acids, and lactic acid. J Dairy Sci 1992; 75: 1725-30

(211) Medish M, Murata S, Frogner K, Fukunaga C, Matsuda L. Glycerol monolaurate in model toxic shock syndrome. In Abstracts of the 91st General Meeting of the American Society for Microbiology, Washington, DC. 1991; B19: 28

(212) Schlievert PM, Deringer JR, Kim MH, Projan SJ, Novick RP. Effect of glycerol monolaurate on bacterial growth and toxin production. Antimicrob Agents Chemother. 1992; 36: 626-631

(213) Wang L, Yang B, Parkin KL, Johnson BA. Inhibition of Listeria monocytogenes by monoacylglycerols synthesized from coconut oil and milkfat by lipase catalyzed glycerolysis. J Agric Food Chem 1993; 41: 1000-05

(214) Petschow BW, Batema RP, Ford LL. Susceptibility of Helicobacter pylori to bactericidal properties of medium-chain

monoglycerides and free fatty acids. Antimicrob Agents Chemother 1996; 40: 302-06

(215) Sprong RC, Hulstein MF, Van der Meer R. High intake of milk fat inhibits intestinal colonization of listeria but not of salmonella in rats. J. Nutr 1999; 129: 1382-89

(216) Bergsson G, Arnfinnsson J, Steingrimsson O, Thormar H. Killing of Gram positive cocci by fatty acids and monoglycerides. Apmis 2001 Oct; 109 (10): 670

(217) Bergsson, Gudmundur, et al, In Vitro Killing of Candida Albicans by Fatty Acids and Monoglycerides. Antimicrob Agents Chemother 2001; 45 (11): 3209-12

(218) Isaacs CE, Thormar H. Membrane-disruptive effect of human milk: inactivation of enveloped viruses. J Infect Dis 1986; 154: 966-71

(219) Isaacs CE, Thormar H. The role of milk-derived antimicrobial lipids as antiviral and antibacterial agents. In: Mestecky J, et al, eds. Immunology of Milk and the Neonate Plenum Press: New York 1991

(220) Coconut oil compound may treat STDs AIDS Patient Care STDS 1999 Sep; 13 (9): 57

(221) Tayag E, Dayrit CS, Santiago EG, et al. Coconut Oil in Health and Disease: Its and Monolaurin's Potential as a Cure for HIV-AIDS, 37th Cocotech Meeting, Chennai, IndiA

(222) Gudmundur Bergsson,1 Jóhann Arnfinnsson,2 Sigfús M. Karlsson,3 Ólafur Steingrímsson,3 and Halldor Thormar. In Vitro Inactivation of Chlamydia trachomatis by Fatty Acids and Monoglycerides. Institute of Biology, University of Iceland, 1 Department of Anatomy, University of Iceland Medical School, 2 and Department of Microbiology, National University Hospital, 3 Reykjavik, Iceland

(223) Abraham RL, Verallo-Rowell VM, Baello BQ. Testing of Lauricidin versus isopropyl alcohol for antisepsis of cutaneous hand microbes to prevent infection. Philippine J Microbiol Infect Dis 2000; 29: 128-35

(224) Abraham RL, Verallo-Rowell VM. Safety and efficacy of monolaurin, a coconut oil extract, versus ethyl alcohol in rinse-

free hand antiseptic gels on healthcare personnel's hands and microbial isolates. J Phil Derm Soc 2001; 10(2): 90-99

(225) Lacson and Verallo-Rowell Case Report: Coconut Oil in a severe case of Oro-genital Herpes Simplex Virus infection. Submitted.

(226) Carpo and Verallo-Rowell:Coconut Oil in a severe case of Hidradenitis Suppurativa. Submitted.

(227) Carpo T, Verallo-Rowell VM. In Vitro Antibacterial Activity of Monolaurin, a Coconut derivative.

(228) Thormar H, Isaacs EC, Brown HR, Barshatzky MR, Pessolano T. Inactivation of enveloped viruses and killing of cells by fatty acids and monoglycerides. Antimicrob Agents Chemother 1987; 31: 27-31

(229) Isaacs CE, Kim KS, Thormar H. Inactivation of enveloped viruses in human bodily fluids by purified lipids. Ann NY Acad Sci 1994; 724: 457-464

(230) Projan SJ, Brown-Skrobot S, Schlievert PM, Vandenesch F, Novick RP. Glycerol monolaurate inhibits the production of ß-lactamase, toxic shock syndrome toxin 1 and other staphylococcal exoproteins by interfering with signal transduction. J Bacteriol 1994; 176: 4204-09

(231) Ruzin A, Novick RP. Equivalence of lauric acid and glycerol monolaurate as inhibitors of signal transduction in staphylococcus aureus. J of Bacteriol 2000; 182: (9): 2668-267

(232) Hormung et al. Lauric acid interferes with viral assembly and maturation. J Gen Virol 1994; 75 (2): 353-61

(233) Nosakhare'Odeh Eghafona. Immune responses following cocktails of inactivated measles vaccine and arachis hypogaea L. (groundnut) or cocos nucifera 1 (coconut) oils adjuvant. Vaccine 1996; 14 (17/18): 1703-06

(234) Witcher K, Novick R, Schcievert P. Modulation of immune cell proliferation by glycerol monolaurate. Clin Diag Lab Immunol 1996 Jan; 10-13

(235) Lim-Navarro PRT, Escobar R, Fabros M, Dayrit CS: Protection effect of coconut oil against E. coli endotoxin shock in rats. Coconuts Today 1994; 11: 9

(236) Ruzin A, Novick RP. Glycerol monolaurate inhibits induction of vancomycin resistance in enterococcus faecalis. J Bacteriol 1998; 180: 182-185

(237) Carpo T, Verallo-Rowell V. oral virgin coconut oil for skin infections (a single-blind randomized active-controlled comparative trial on the efficacy of oral virgin coconut oil vs. Cloxacillin for localized bacterial skin infection—a pilot study). On-going study.

(238) Byrnes S. I've got a Lovely Bunch of Coconuts: Coconut Holds Promise for the Immune Suppressed People. Power Health, 1-7 N Engl J Med 1994; 330: 12

(239) Lehmann H, Robinson K, Andrews J, et al. Acne therapy: A methologic review. J Am Acad Dermatol Aug 2002.

(240) Thiboutot D. Acne: 1991-2001. J Am Acad Dermatol 2002 July; 47: 1

(241) Pineda-DLS JP, Verallo—Rowell VM, Open study on the safety and efficacy of 2% monolaurin gel in the treatment of mild to moderate inflammatory facial acne vulgaris Submitted. 2004.

(242) A Vehicle-Controlled Trial on the Safety and Efficacy of 2 % Monolaurin Gel in the Treatment of Mild to Moderate Inflammatory Facial Acne Vulgaris 2004/ Submitted.

(243) Back O, Faergemann J, Hornqvist R. Pityrosporum folliculitis: a common disease of the young and middle-aged. J Am Acad Dermatol 1985 Jan; 12 (1 Pt 1): 56-61

(244) Jacinto-Jamora S, Tamesis J, Katigbak ML. Pityrosporum folliculitis in the Philippines: diagnosis, prevalence, and management. J Am Acad Dermatol 1991 May; 24 (5 Pt. 1): 693-96

(245) Schmidt A. Malassezia furfur: a fungus belonging to the physiological skin flora and its relevance in skin disorders. Cutis 1997 Jan; 59 (1): 21-4

(246) Verallo-Rowell V. Personal communication on clinical experience with monolaurin gel in the treatment of Pityrosporum folliculitis June 2003

(247) Dela Cruz MDC, Verallo-Rowell VM. A double blind randomized controlled trial to compare the efficacy and safety of 2% coconut monoglyceride gel with 2% monolaurin gel

and 2% ketoconazole lotion in the treatment of pityrosporum folliculitis among adolescents and middle aged adults

(248) Sarte KAR, Verallo-Rowell VM. A double blind randomized control trial to compare efficacy and safety of 2% coconut monoglyceride gel with 2% monolaurin gel and 2% ketoconazole lotion in the treatment of Pityrosporum folliculitis among adolescents and middle aged adults

(249) Agero AL, Verallo-Rowell VM. A randomized double-blind controlled trial comparing extra virgin coconut oil with mineral oil as a moisturizer for mild to moderate xerosis. J Contact Derm 2004 J Contact Dermatitis 15; (3)September 109-116

(250) Rele AS, Mohile RB. Effect of mineral oil, sunflower oil, and coconut oil on prevention of hair damage. J Cosmet Sci 2003 Mar-Apr; 54 (2): 175-92

(251) National Center for Complementary and Alternative Medicine. [accessed 2003 Sept. 09] URL:http://nccam.nih.gov/research

(252) Elias PM. The epidermal permeability barrier: from the early days at Harvard to emerging concepts. J Invest Dermatol 2004; 122: 36-9

(253) Sator PG, Schmidt JB, Honigsmann H. Comparison of epidermal hydration and skin surface lipids in healthy individuals and in patients with atopic dermatitis. J Am Acad Dermatol 2003; 148: 35

(254) Lowe NJ, Friedlander J. Sunscreens: rationale for use to reduce photodamage and phototoxicity. In: Lowe NJ, Shaath NA, Pathak MA,eds. Sunscreens: Development, Evaluation, and Regulatory Aspects. Second Ed. New York: Marcel Dekker, Inc, 1997: 38

(255) Oxygen species and antioxidant protection in photodermatology. In: Nicholas Lowe, Nadim Shaath, Mudhu Pathak eds. Sunscreens: Development, Evaluation, And Regulatory Aspects, Marcel Dekker Inc: New York

(256) Darr D, Pinnell SR. Reactive Trevithick J et al. Topical tocopherol acetate reduces post-UVB, sunburn-associated erythema, edema and skin sensitivity in hairless mice. Arch Biochem Biophys 1992; 196: 575-82

(257) Henry GE, Momin RA, Nair MG, Dewitt DL. Antioxidant and cyclooxygenase activities of fatty acids found in food. J Agric Food Chem 2002; 50: 2231-34

(258) Elmets C, Singh D, Tubesing K, Matsui M, Katiyar S, Mukhtar H. Cutaneous Photoprotection from ultraviolet injury by green tea polyphenols. J Am Acad Dermatol 2001; 44: 425-32

(259) Tobi SE, Gilbert M, Paul N, McMillan TJ. The green tea polyphenol, epigallocatechin-3-gallate, protects against oxidative cellular and genotoxic damage of UVA radiation. Int J Cancer 2002; 102 (5): 439-44

(260) Verallo-Rowell VM, Nograles KB, Yu NT. Chemoprotective effect of green tea extract on UV-irradiated Asian skin. Presented: Korean Dermatologic Society Photomedicine Meeting. June 9, 2004, Seoul, Korea

(261) Verallo-Rowell VM, Arevalo C, Chong C, Cuaso-Cruz J, Nograles K, Sanchez E. Topical antioxidants: randomized double blind study with Filipino skin irradiated with UVA, UVB, Infrared Light. Poster Presentation at the American Academy of Dermatology Annual Meeting, New Orleans. February 18-22, 2005

(262) Yu NJT, Nograles KB, Verallo-Rowell VM. Determination of photoprotective and antioxidant properties of topical virgin coconut oil on Filipino skin irradiated with solar simulated light. Submitted.

(263) Coconut water. 19 Citations http://www/ncbi.nim.nih.gov. accessed 20 Mar 2005

(264) Van Oberbeek J, Siu R, Haagan-Smit A. Factors in coconut milk essential for growth and development of very young Datura embryos. Science 1941; 94: 350

(265) Mauney JR, Hillman WS, Miller CO, Skoog F, Clayton RA, Strong FM. Bioassay, purification and properties of a growth factor from coconut. Physiol Plant. 1952; 5: 485-97

(266) Radley M, Dear E. Occurrence of gibberellin-like substances in the coconut. Nature 1958; 182 (4642): 1098

(267) Van Staden J, Drewer SE. Identification of cell division inducing compounds from coconut milk. Physiol Plant. 1974; 32: 347-35
(268) Mamaril JC, Paner ET, Trinidad LC, Palacpac ES, C dela C. Enhancement of seedling growth with extracts from coconut water. Philip J Crop Sci 1988; 139 (1): 1-7
(269) Uichangco N. Studies on the chemical changes of coconut water of matured nuts during storage. Unpublished thesis, University of the Philippines at Los Banos 1969
(270) Yson RL. Study of the Gibberellin level of coconut water from matured nuts during storage. Unpublished thesis, University of the Philippines at Los Banos 1994
(271) Miller CO, Skoog F, Von Saltza MH, Strong FM. Kinetin, a cell division factor from deoxyribonucleic acid. J. Am. Chem. Soc 1955; 77: 1392
(272) Miller CO, Skoog F, Okumura FS, Von Saltza MH, Strong FM. Isolation, structure and synthesis of kinetin, a substance promoting cell division. J Am Chem Soc 1956; 78, 1375-80
(273) Barciszewski J, Siboska GE, Pedersen BO, Clark BFC, Rattan SIS. Furfural, a precursor of the cytokinin hormone kinetin, and base propenals are formed by hydroxyl radical damage of DNA. Biochem Biophys Res Commun 1997; 238:317-19
(274) Barciszewski J, Rattan SIS, Siboska G, Clark BFC. Kinetin—45years on. Plant Sci. 1999;148: 37-45
(275) Barciszweski J, Siboska G, Rattan SIS, Clark BFC. Occurrence, biosynthesis and properties of kinetin (N6-furfuryladenine). Plant Growth Regulation 2000; 32: 257-65
(276) Sharma SP, Kaur P, Rattan SIS. Plant growth hormone kinetin delays ageing, prolongs the lifespan and slows down development of the fruitfly Zaprionus paravittiger. Biochem Biophys Res Commun 1995; 216: 1067-71
(277) Sharma SP, Kaur J, Rattan SIS. Increased longevity of kinetin-fed Zaprionus fruitflies is accompanied by their reduced fecundity and enhanced catalase activity. Biochem Mol Biol Int 1997; 41: 869-75

(278) Rattan SIS, Clark BFC. Kinetin delays the onset of ageing characteristics in human fibroblasts. Biochem Res Commun 1994; 201: 665-7

(279) Weinstein GD, McCullough JL, Ali NN, Shull TF, Moudy D: A double-blind vehicle-controlled study of kinetin lotions for improving the appearance of aging photodamaged facial skin with 24 weeks of twice daily topical application

(280) Weiss JS, Ellis CN, Goldfarb MT, Voorhees JJ: Tretinoin treatment of photodamaged skin. Cosmesis through medical therapy. Dermatol Clin 1991; 9: 123-9

(281) Olsen A, Siboska GE, Clark BFC, Rattan SIS. 1999: N6-furfuryladenine, kinetin, protects against Fenton reaction—mediated oxidative damge to DNA. Biochem. Biophys. Res. Commun. 265, 499-502

(282) Verbeke P, Siboska GE, Clark BFC, and Rattan SIS. 2000: Kinetin inhibits protein oxidation and glyoxidation in vitro.Biochm biophys Res Commun. 276, 1265-1270.

(283) Rochat A, Kobayashi K, Barrandon Y. Location of stem cells of human hair follicles by clonal analysis. Cell 1994; 76: 1063-73

(284) Jahoda CA, Reynolds AJ. Dermal—epidermal interactions—follicle-derived cell populations in the study of hair-growth mechanisms. J Invest Dermatol 1993; 101 (1): 33S-38S

(285) Akiyama M, Smith LT, Holbrook KA. Growth factor and growth factor receptor localization in the hair follicle bulge and associated tissue in human fetus. J Invest Dermatol 1996; 106: 391-6

(286) Harmon CS, Nevins TD. IL-1 alpha inhibits human hair follicle growth and hair fiber production in whole organ cultures. Lymphokine Cytokine Res 1993; 12: 197-203.

(287) Herbert JM, Rosenquis T, Gotz J et al. FGF 5 as a regulator of the hair growth cycle: evidence from targeted and spontaneous mutations. Cell. 1994; 78: 1017-25.

(288) Du Cros DL. Fibroblast growth factor and epidermal growth factor in hair development. J Invest Dermatol 1993; 101(1 Suppl): 106S-113S.

(289) DM, Ring BD, Yanagihara D, et al. Keratinocyte growth factor

is an important endogenous mediator of hair follicle growth, development, and differentiation. Normalization of the nu/nu follicular differentiation defect and amelioration of chemotherapy-induced alopecia. Am J Pathol 1995; 147: 145-54

(290) Batch JA, Mercuri FA, Werther GA. Identification and localization of insulin-like growth factor-binding protein (IGFBP) messenger RNAs in human hair follicle dermal papilla. J Invest Dermatol 1996; 106: 471-75

(291) Paus R, Heinzelmann T, Schultz KD, et al. Hair growth induction by substance P. Lab Invest 1994; 71: 134-40

(292) Sawaya ME. Purification of androgen receptors in human sebocytes and hair. J Invest Dermatol 1992; 98 (6 Suppl): 925-65

(293) Itami S, Kurata S, Sonoda T, et al. Interaction between dermal papilla cells and follicular epithelial cells in vitro: effect of androgen. Br J Dermatol 1995; 132: 527

(294) Espino TM et al. Production of virgin coconut oil and lipase catalyzed synthesis of ß-monoglyceride. Submitted,2004.

(295) Boceta NM, De Leon SY, Masa DB Contributing authors and editors. In 3Rs on Coconut Flour (Philippine Coconut Authority publisher)

(296) Brand-Miller J, Wolever TMS, Foster-Powell K, Colagiuri C. The New Glucose Revolution The Authoritative Guide to the Glycemic Index, the Dietary solution for lifelong health. 2004 Muze Inc

(297) Trinidad TP, Valdez DH, Loyola AS, Mallillin AC, Askali FC, Castillo JC, Masa DB. Glycaemic index of different coconut (cocos nucifera) flour products in normal and diabetic subjects. Br J Nutr 2003; 90: 551-55

(298) Trinidad TP, Loyola AS, Mallillin AC, Valdez DH, Askali FC, Castillo JC, Resaba RL, Masa DB. The cholesterol-lowering effect of coconut flakes in humans with moderately rasied serum cholesterol. J Med Food 2004; 7 (2): 136-40

(299) Nutrition and Your Health: Dietary Guidelines for Americans 6th Edition 2005

(300) Nutrition Source Harvard School of Public health http://www.hsph.harvard.edu/now/

BVG